Studies in Social and Cultural Anthropology

Kulturstiftung Sibirien

Kirill V. Istomin and Mark J. Dwyer

Reindeer herders' thinking

A comparative research of relations between economy, cognition and way of life

Verlag der Kulturstiftung Sibirien
SEC Publications

ЕВРОПЕЙСКИЙ
УНИВЕРСИТЕТ
В САНКТ-ПЕТЕРБУРГЕ

This publication has been made possible with the assistance of the Center for Arctic Social Studies, European University at St. Petersburg

Bibliografische Informationen der Deutschen Nationalbibliothek:
Die Deutsche Nationalbibliothek verzeichnet diese Publikation in der Deutschen Nationalbibliografie: detaillierte bibliografische Daten sind im Internet über <http://dnb.d-nb.de> abrufbar.

Cover photo:
Mark Dwyer

Typesetting:
Kulturstiftung Sibirien gGmbH, Fürstenberg/Havel

Printer:
Books on Demand GmbH, Norderstedt

Electronic edition:
https://dh-north.org/publikationen/reindeer-herders-thinking/en

ISBN: 978-3-942883-73-3

Printed in Germany

CONTENTS

Summary

This book is based on more than a decade of anthropological fieldwork and scholarship among Komi and Nenets nomadic reindeer herders of North-eastern Europe and North-western Siberia. Focused on herding techniques and the way of life of arctic nomads, the authors cogitate the unique attributes of reindeer herding and how these influence the herders' cognitive skills. Two central cognitive abilities are explored: the ability to "find their way" in expansive and homogenous arctic tundra terrain, often in extreme weather conditions and navigating with neither maps nor navigation equipment, and the ability to "decipher and predict" reindeer behaviour. This book acknowledges and reviews current theories and models of human cognition developed in cognitive science. The authors' build bridges between cognitive science and anthropology by presenting further case studies that reveal and "demystify" cognitive mechanisms. Axiomatically, they challenge the field of anthropology by demonstrating fundamental weaknesses and debunking anthropological theories that ignore cognitive facts. The authors advocate that the field of anthropology should no longer isolate itself from other scientific disciplines, since, in doing so, its marginalisation will amplify and its relevance diminish. This book exemplifies the contribution of anthropology to building greater understanding of human cognition. However, this can only be achieved through embracing advancements made in other disciplines rather than ignoring their existence.

Kirill V. Istomin is a professional anthropologist and a senior researcher at the Center for Arctic Social Studies, European University at St. Petersburg (St. Petersburg, Russia) and at the Institute of Language, Literature and History, Komi Science Center of the Ural Division, Russian Academy of Science (Syktyvkar, Komi Republic, Russia). Istomin worked at the Max Plank Institute for Social Anthropology (Halle, Germany) for 7 years and has long and deep experience conducting fieldwork among his native Komi (whose language he has spoken since childhood) and Nenets reindeer herders. His areas of scientific interests are broadly those contained in this book. Istomin is married, has two children as well as a passion for reindeer herding, literature and life in the north.

Mark J. Dwyer conducted his multidisciplinary doctoral research at the University of Cambridge (Scott polar research Institute & Selwyn College). Supervised by Piers Vitebsky (anthropology) and Gareth Rees (physicist), as well as an external supervisor, and fellow Selwyn graduate, Nicholas Tyler from Trømsø University (biologist), Dwyer subsequently conducted postdoctoral research at Cambridge's Geography Department before entering the field of international development. Specialising in public-private infrastructure and finance for over a decade in Africa, Dwyer returned to the north and is currently CEO of a group of companies operating in the fields of mining exploration, energy and rapid deployment. Dwyer is married to a former Komi reindeer herder, has three children and is passionate about reindeer herding, oral history, art and the arduous yet indelible connections with the land.

Preface

This book represents the culmination of over a decade of both collaborative and independent research carried out by Istomin and Dwyer. The manuscript was first drafted between 2011 and 2012, however, its publication was delayed owing to the authors' career commitments. We are, therefore, grateful to the Foundation for Siberian Cutures SEC for accepting to publish our works.

Publishing a book that was written almost ten years ago, provides us with a unique opportunity to assess how the evidence, ideas, beliefs, as well as the hopes that we had for the manuscript in 2012, may be viewed today. We thus consider it important to provide a brief review of relevant research carried out over the past decade, which is absent in the body of our book.

Our updated literature review, revealed that – of the two main topics addressed in this book – namely spatial orientation and human-reindeer interaction within a coupled human-animal nomadic system – the later area of research received the most attention by anthropologists over the past ten years. We particularly wish to mention the works by Charles Stépanoff and colleagues (Stépanoff 2012, 2017; Stépanoff et al. 2017; Stépanoff & Vigne 2018), Alex Oehler (Oehler 2020), David Anderson (Anderson et al. 2017) and Vladimir Davydov (Davydov 2017) with regards to research undertaken on human-animal relations in a reindeer herding system. Many of these academics developed similar ideas to those we propose in our "dynamic mutual adaptation model", which we discuss in chapter five (see also Istomin & Dwyer 2010). We are pleased to report that many of them have developed the descriptions and concepts of "dynamic mutual adaptation" much further than we could do ten years ago. were somewhat developed and the spheres where these ideas have been applied by far exceed everything we either thought about. However, we believe that our research on mechanisms of perceiving deciphering and predicting reindeer behaviour that we present in chapter seven, remains relevant and novel to this day.

Conversely, the study of spatial perception and spatial orientation among reindeer herding nomads, the area of research we consider to be central to our work, received relatively little attention during the past ten years. One can indeed only point to the publications of Vladimir Adaev (Adaev 2015a, 2015b, 2015c, 2015d, 2016, 2018; Adaev & Rahimov 2015) on spatial orientation among Nenets. These works (all of which are in the Russian language) are descriptive and void of analysis. We are miffed by many of the above publications, since, despite referring to similar groups of Nenets people, the ethnographic descriptions produced by Adaev often conflict with our findings. Adaev, in particular, reports that Taz Nenets make regular use of cardinal directions in their "spatial orientation" and always know the direction of the north. Adaev reports that Taz Nenets make use of astronomic reference points – including the sun, moon, stars and planets. We are surprised by these differences and cannot explain them. We are equally astonished that while Adaev cited our work approvingly (Adaev 2015b: 95, 2018: 89) the author failed to recognise the existence of conflicting accounts let alone explain them. Furthermore, Adaev writes that while Nenets people are always aware of the position of north and south, they often do not know the position of west and east. We question such a partial use of cardinal directions: in our opinion anyone who knows the position of north and south cannot fail to know the position of east and west if he/she understands these concepts (and Adaev claims that Nenets have such knowledge). In addition, Adaev reports that the Nenets rely mostly on the moon rather than the sun as a means of orientation, including for taking cardinal directions, which we also find difficult to grasp. In any case, the above claims are so contrary to our findings that they certainly warrant further investigation.

Adaev aside, we note that Valeria Vasilyeva (Vasilyeva 2018), on the basis of her fieldwork among the Dolgan people of the Taimyr Peninsula, reached a number of similar conclusions to the ones made in this book. This particularly concerns the reliance of the tundra nomads' use of "mental maps" rather than "route following" (vistas), the regionalised structure (element) of the mental maps and the nomads' dependence of mental map structures on their means of and purposes for traveling.

We conclude from our updated review of the literature that most of our findings and conclusions remain current and relevant. We in fact observe that the anthropology and cognitive science nexus, which was a popular area of research at the beginning of the present century, and which continues with the publication of this book, appears to have been relatively short lived. This does not mean that interest in how culture influences human thinking (and, at a more basic level, the human brain) has diminished. To the contrary, interest seems to be stronger today than it was ten years ago. However, anthropology, a discipline in which, one would expect, such research to take place, has broadly withdrawn from cognitive studies and handed its baton to cross-cultural psychology, cultural neuroscience and other branches of empirical positivistic disciplines to advance our understanding in this area. We find such intellectual and "disciplinary isolationism" regretful and hope that this book may serve to demonstrate that anthropology has a place to advance our understanding of human cognition, as well as other areas of study that require collaborative and multidisciplinary scholarship.

Acknowledgements

We would like to thank the many people, colleagues and friends, who contributed to our work at different stages. However, we first and foremost wish to thank, and also celebrate, the Komi and Nenets people who were our *raison d'être*. We feel humbled to have had the privilege of documenting, studying and interpreting their cultures, which we have done in good faith and to the best of our ability.

We wish to thank Dr. Brian Donahoe, who edited the first draft of this book. We acknowledge **Prof. Dr. Joachim Otto Habeck, Dr. Nicholas Tyler,** Prof. Dr. Günther Schlee, Prof. Dr. Nikolai Vakhtin and Dr. Erich Kasten, whose insights helped us get our book through to publication. We would also like to express our gratitude to our colleagues **Dr. Piers Vitebsky, Dr. Gareth Rees,** Prof. Tim Bayliss-Smith, Dr. Elena Liarskaia, Dr. Vladislava Vladimirova, Dr. Virginie Vate, Dr. Andrea Bender, Dr. Ludeck Broz, as well as to the members of the former CLLP working group in the **Max Planck Institute for Social Anthropology, the Scott Polar Research Institute and the Department of Geography (University of Cambridge) Selwyn College (University of Cambridge)** for their fruitful and patient insights and/or support that are reflected in numerous parts of this book and made our work possible.

We would also like to thank our many friends and colleagues from Russia, particularly **Prof. Dr. Yuri Shabaev** (Komi Science Center, Russian Academy of Science), **Prof. Dr. Andrei Golovnev** (Ural Division, Russian Academy of Science), Prof. Dr. Dmitri Funk (Institute of Ethnology and Anthropology, Russian Academy of Science) for their advice, as well as A. Makarchuk (Bol'shaya Inta Sovkhoz), A. Kaplun (Tazovsky Agricultural Cooperative), families of A. Shurakov and D. Alekseevich whose assistance made our fieldwork possible. We wish to acknowledge the substantial contribution of the late Mrs. Eileen J. Dwyer who spent countless hours working with the authors for over more than a decade on many versions of manuscripts – her work ethic, sharp mind and attention to detail was invaluable. Our deepest gratitude, however, should be addressed to our special Komi and Nenets friends, in particular, the Yangasov, Larionov, Filippov, Kanev and Semyashkin extended families (Komi) and the Salinder, Khudi, and Tyseda extended families (Nenets). We were instantly accepted as "family" and this book was only possible thanks to their guidance, assistance, patience, and extraordinary generosity, which was truly eye opening and life changing for both authors.

Finally, we wish to show our gratitude for the financial support of the Deutsche Forschungsgemeinschaft (project E86 of the Collaborative Research Centre (SFB) 586 "Difference and Integration"); the Natural Environment Research Council (UK), Social & Economic Research Council (UK), the Scott Polar Research Institute (University of Cambridge, UK), the Department of Geography (University of Cambridge, UK) and the wonderful Selwyn College (University of Cambridge, UK).

1 INTRODUCTION

1.1 Three flags to be raised

This book is about nomadic reindeer herding. It is written by professional cultural anthropologists with multidisciplinary expertise and training in physical sciences (Dwyer) who have spent their entire academic careers studying it. However, this book, we feel, is rather different both in its aims and in its approach from most anthropological works that have thus far been written on this topic. Sometimes these differences are so pronounced that some of our colleagues, we are afraid, would hesitate to regard the studies presented in the chapters that follow as being genuinely considered anthropological at all. Anticipating this reaction, we would like to start this book by raising several flags that mark out our position vis-à-vis the aim and mission of anthropology as a discipline.

To raise the first flag, we are among those who believe that the ultimate aim of anthropology is to *explain* cultural phenomena rather than simply to describe and interpret them. Therefore, we reject the still widespread view that anthropology is nothing more than "cultural critique" (Marcus and Fischer 1983). The main reason for our rejection of this position is our deep concern, increasingly shared by many of our colleagues, about the present state and the future of our discipline. Indeed, our discipline started about 100 years ago as an enterprise aimed at describing and explaining the variation in the ways of life, social organisation, and spiritual lives of different groups of people populating our planet. Most of our colleagues would agree, however, that, despite the enormous body of empirical material that has been collected, anthropology has not made much progress in reaching this goal. To the contrary, for the last quarter of a century anthropological research has been predominantly interpretive, thereby defying theoretical assessment. As a result, cultural anthropology is still in the fragmented state of a "pre-paradigmatic" science, to use the term introduced by Kuhn (1962). Furthermore, the widespread retreat of anthropology into interpretivism precludes, we are concerned, any possibility of the discipline ever reaching a paradigmatic level. It follows, therefore, that if anthropologists continue to consciously position the discipline as one that does not know anything for certain and, on occasion, take pride in this fact, colleagues from other disciplines, policy makers, and lay people will simply cease to take anthropology seriously. In fact, this may already be the case.

Our outcry is not intended to imply that interpretations are unnecessary or that interpretive research should be banished. Instead, we simply want to assert that anthropologists should not lose sight of the ultimate aim: to learn about the plethora of human cultures and ways of life in a manner that is amenable to testable theoretical generalisations. To achieve this – and we now come to the second flag we wish to raise – we need interdisciplinary cooperation. In order to unravel, we believe, we should not tolerate, let alone consciously build, walls that currently separate our discipline from other disciplines also attempting to explain human nature, thinking, and behaviour. Of course, anthropology's counter-positivistic retreat has gone so far that we sometimes misunderstand each other. In fact, for some of our colleagues, it is often easier to grasp a hunter from the African savannah than a fellow scientist working in the natural sciences. However, our conviction is that we should speak and work together, not only because we have a great deal to learn from one another but also because advancing understanding of the *human condition* is only possible with such dialogue.

Finally, to raise the third flag, we are among those who take seriously the definition of a society's culture, as proposed by Ward Goodenough, seriously: "a society's culture consists of whatever it is one has to know or believe in order to behave in a manner acceptable to its members" (1957, 168). Two aspects of this definition seem especially important to us. First of all, it stresses a rather simple but frequently forgotten fact that culture manifests itself as a set of behavioural patterns. In other words, for culture, as for any social phenomenon, to exist means to shape human behaviour (Engels 1907). Secondly, the definition points out that culture shapes behaviour through shaping beliefs, attitudes, inferences, and, we would add, other demonstrative aspects of individuals' cognition. Culture is, therefore, essentially a cognitive phenomenon. This viewpoint is fully compatible with the fact that culture is also a social phenomenon – as has been well known, at least since the time of Vygotsky (1978 [1934]): individual cognition develops through and is essentially shaped by social interactions. It is, therefore, through social interactions with their specific behavioural patterns that culture as a cognitive phenomenon both manifests and reproduces itself. On the other hand, to state that culture "resides" or "exists" in these interactions is either mysticism or a terminological extravagance. When Clifford Geertz writes that culture is not "happening in the head" but that the "natural habitat" of culture (as well as language and human thinking in general) "is the house yard, the marketplace, and the town square," (Geertz 1973: 45) he is, in our opinion, employing a metaphor. Indeed, regarding language, what really exists "in the market place" and "in the streets" is strings of sounds, acoustic waves with various characteristics, and it is only the perception of these waves by humans and the further processing of these percepts in their brains that make up language. Similarly, culture exists in the minds of people who interact, while interaction itself represents nothing more than a behavioural sequence that is turned into a set of actions with certain social or cultural significance within our minds.

These three theoretical and methodological statements inform our view of what a study of cultural phenomena represents – and by extension, what an anthropological study – should be. We define such a study as an explanatory inquiry into cognitive mechanisms that give rise to certain sustained forms of behaviour that are specific to certain groups of people. The groups in question can be ethnic, religious, professional, or other. The cognitive mechanisms can include beliefs, dispositions, desires, and other mental representations, as well as forms of perception, inference and memory. The explanatory character of the inquiry includes formulating hypotheses about what these cognitive mechanisms are and how they work to produce sustained forms of behaviour. This infers identifying how these mechanisms emerge and transform during the course of a group's history, how they get incorporated into individual cognitions of group members (throughout their lives) and how they are modified in situations of unprecedented change in the social and natural environment. In addition, these hypotheses should lead to predictions that can be tested empirically. Both the formulation of these hypotheses and their testing can and should be informed by theoretical and methodological advancements in disciplines other than in anthropology alone: cognitive psychology can be especially relevant here, as can linguistics, certain biological disciplines, such as physical anthropology, ecology and ethnology. The so-called standard methods of ethnographic inquiry – participant observation and ethnographic interviews – are useful tools for anthropological inquiry. However, these methods are seldom sufficient to complete a rigourous study. As such, this view constitutes our most radical departure from anthropological inquiry as it is commonly understood: indeed, it is customary to think that the aim of such an inquiry is to produce a description of "how things are" from the viewpoint of the members of a certain group with, it seems to us, an implicit (and sometimes explicit) belief that such a description constitutes an explanation of why these members "do what they do." In our opinion, however, it is extremely interesting, important and, furthermore, necessary to explain why, from the viewpoint of the informants, "things are the way they are." Only after such an explanation is provided, can we speak, with any level of accuracy, of understanding. The studies in this book have been designed and carried out in accordance with the approach outlined above.

1.2 Object and aims of this book

This book represents our attempt to reply to arguably the most basic question in cross-cultural psychology and cognitive anthropology: How, if at all, can culture influence human thinking beyond the level of its content? In other words, we wish to understand if culture is capable of influencing not only what people think, but also how they think (i.e., the kinds of cognitive processes that support their thinking) and, if it can, then how deep can this influence be and through what mechanisms is it effected. In order to find answers to these questions, we will focus on the cognitive phenomena behind the specific behaviour patterns that constitute what is commonly referred to as nomadic reindeer herding among indigenous inhabitants of northeastern Europe and western Siberia. In short, we seek an answer (partially, naturally) to the following concrete research question:

> *What must one know and believe, and how must one think, in order to be a nomadic reindeer herder?*

This choice of research question has been informed by our understanding of the aim and purpose of anthropological inquiry as outlined in the first section of this chapter. Following this understanding, we will try to answer these questions in a way that not only describes the cognitive phenomena related to the practice of reindeer herding, but also explain why these phenomena exist.

Needless to say, we cannot provide an all-encompassing treatment of "reindeer herding cognition" in the framework of a single volume. Therefore, the studies contained in this volume focus on two domains of cognition that are related to solving two types of cognitive tasks routinely faced by reindeer herders. The first type of tasks is related to planning and performing spatial actions and includes such problems as the planning of movement and navigation. The cognitive mechanisms responsible for solving this type of tasks include (but are not limited to) perceiving and representing the landscape and the spatial layout of its elements, and cognitive operations of position fixing, course inference and course keeping. These usually fall under the rubric "spatial cognition" (Downs and Stea 1973). The second type of tasks is related to perceiving, deciphering, predicting and influencing animal behaviour. Although the cognitive mechanisms responsible for solving these tasks have thus far rarely been systematically researched, we will attempt to prove that, at least among reindeer herders, they are closely related to the cognitive domain usually referred to as Theory of Mind (Carruthers and Smith 1996; Dennett 1981; Leslie 1987). The choice of these two types of tasks, as the primary focus of our research on the cognition of nomadic reindeer herding pastoralists, is quite logical. Indeed, the first type of tasks should logically be the most prominent for nomads. Conversely, it is the second type of tasks that one would expect to be the most prominent for pastoralists. Hence, the influence, of a nomadic-pastoralist culture and way of life, on cognition can be expected to reside primarily in the domains related to these tasks. This makes them an obvious object of interest as far as general questions about the relation between culture and cognition are concerned.

1.3 Culture, cognition and the theoretical approach of this book

Addressing the research questions outlined in the previous section demands a theoretical framework capable of informing a fruitful research approach and design. A brief inquiry into theoretical positions frequently taken with respect to the relation between cultural and cognitive phenomena would be a logical first step to formulating such a framework.

The two most extreme of these positions can be referred to as psychocultural relativism and psychological absolutism (terms adapted from Berry et al. 2002). According to the former, human cognition – both its form and content – is entirely shaped by culture: no cognitive universals can be assumed to exist, and representatives of different cultures can and should be expected to have qualitatively different forms of cognition. An obvious methodological consequence of this point of view would be that theoretical models and accounts of cognitive processes obtained from studies using representatives of one culture (let's say American) cannot be used to make sense of cognitive and behavioural phenomena observed in other cultures, even if those phenomena appear to be very similar. Thus, we cannot assume that representatives of other cultures have the same perceptual mechanisms that Americans have (that they perceive objects in three dimensions, for example), that their perception is guided by selective attention, that they use percepts to build up concepts, that they store their percepts and concepts in the same two types of long-term memory (semantic and episodic) and retrieve them by the means of situation-bounded mechanisms, and so on. In contrast, according to psychological absolutism, forms of cognition are naturally[1] given and, therefore, essentially similar across our species. For this position, an essential distinction is presumed to exist between cognitive processes and cognitive content (Nisbett 2003): while the content is expected to vary across cultures, the cognitive processes are assumed to be the same among the representatives of all cultures. In other words, reindeer herders of the tundra and European office managers are expected to exhibit similar processes of perception guided by selective attention, the formation of representations, memory, inference, and so forth, and the only thing that differs is that the former apply these processes to reindeer and reindeer herding, while the latter apply them to other objects and tasks present in their immediate environment.

A curious observation about these extreme positions is that one of them is often explicitly claimed to be right, but almost never assumed in actual research, while the other is often implicitly assumed, but never explicitly claimed to be right. Indeed, although the absolute majority of studies in psychology and cognitive science are performed on participants recruited from modern industrialised societies, they are still commonly alleged to contribute to an overall understanding of human cognition by discovering its general principles. This claim, of course, can be made only if psychological absolutism is implicitly assumed. However, we are not aware of any explicit claims made by any psychologists or cognitive scientists to the tune that cognitive processes must be similar, from their details up through all levels of cognition and

1 We are reluctant to write "genetically" here because genetic inheritance is a rather tricky concept as far as complicated phenomena such as cognition are concerned. To state that something is "naturally" given does not necessarily imply that it is coded in genes in any simple way.

across all cultures. Indeed, the empirical results obtained in the recently developed field of cross-cultural psychology indicate that such a claim would simply be wrong (see, for example, Berry et al. 2002 for a comprehensive review). In addition, psychocultural relativism is very much consistent with the general relativist stance that is commonly believed to be the necessary starting point for any cross-cultural research. Therefore, since the time of Herskovits (Herskovits 1948), this position has often been recommended. If not always as an accurate description of reality, it is at least a valid and justified methodological and theoretical starting point – especially within anthropology and ethnopsychology (Allwood 1998; Kim et al. 2006; Shweder 1991). However, we do not see any indication that this recommendation has ever been taken seriously in any empirical research performed in these disciplines. Indeed, as anthropologists, we have read a number of works in our discipline that describe how representatives of different cultures have very different concepts of various material, social, and spiritual phenomena: feel different emotions, make different judgments and entertain different beliefs about them, attend to them differently, and form different desires about their current and prospective configurations. Very often these descriptions are used to clarify the meaning of the phenomena to the people about whom the research was conducted. However, none of these works asks whether these people actually form representations and concepts as a part of their cognitive processes, if these can be meaningfully perceived and classified as beliefs, feelings and/or dispositions (desires), if they indeed perform operations based on these representations of the sort we would call judgments, or indeed if the whole concept of meaning can be applied to their cognitive processes at all. Nevertheless, these questions are exactly those that need to be asked first if psychocultural relativism is to be accepted as the theoretical starting point.[2] We believe that the refusal to ask them is quite justified: since behavioural parallelism is not sufficient to assume psychological parallelism in accordance with the classical formulation of the position (one cannot assume that people in other cultures have beliefs, meanings, or judgments just because they behave as if they do), and since one cannot observe cognitive processes directly, these questions simply cannot be assumed. Therefore, anthropology, if it indeed were based on the psychocultural-relativistic assumption, would have to become a purely behavioural science, a kind of human ethology not dissimilar to animal ethology. Indeed, just as an ethologist would strongly oppose the conclusion that a dog that comes and puts its head in your lap for a stroke has "friendly feelings" or is "looking for compassion" (because this conclusion is anthropomorphic – Kennedy 1992), an ethnologist who subscribes to psychocultural relativism, at least theoretically, should strongly oppose any interpretation

2 For example, some of these questions have been recently asked by Tim Ingold (Ingold 2000). However, not in relation to concrete empirical research and not in the way psychocultural relativism would suggest. Instead, what Ingold attempted to demonstrate is that cognitive models assumed in "classic" cognitive science can provide poor or only partial descriptions of how modern industrialised and people with a subsistence way-of-life "really think."

of a behaviour of a representative of another culture in terms of analogous feelings, desires, or any other familiar cognitive phenomena from the researcher's own cultural repertoire (because this interpretation is ethnocentric). The whole point and strength of anthropology, however, rests in the exact opposite approach: in its ability to go beyond mere behavioural descriptions to the realm of feelings, desires, and hidden configurations,[3] to provide, in other words, the "thick description" that Geertz (1973) has called for. This approach has so far proven to be an effective means of explaining the behaviour of representatives of other cultures. However, for this approach to work, a certain degree of psychological unity must be assumed to exist across cultures. This mere suggestion infers that psychocultural relativism is wrong.

The conclusion that should be reached from this discussion is that both psychological absolutism and psychocultural relativism (despite the acclaim given to it by some anthropological theoreticians) are most likely incorrect ontological positions and sterile epistemologically. A suitable starting point for an investigation into culturally specific cognitive phenomena should lie in between these two extremes, within the family of theoretical positions that is frequently referred to as psychocultural universalism (Berry et al. 2002). All the positions in this family share the fundamental assumption that basic cognitive processes such as perception, representation, memory, inference, and so forth are universal across the human species. That means that we are not likely to find a qualitatively distinctive cognitive process present in some cultural groups that are absent in other groups (Berry et al. 2002). At the same time, culture can and does influence cognitive processes beyond the level of their content: there exist culturally specific *forms of cognitive processes* (culturally specific cognitive mechanisms), not merely culturally specific *contents of cognition*. The best way to think about these forms, however, is to treat them as group-specific variations of cognitive universals. As John Berry described this, "culture plays different variations" on "a set of psychological givens" that are "common to all members of the species" (Berry 2004: 6–7). This ontological position has an important epistemological implication: because basic cognitive processes are assumed to be universal, behavioural parallelism can be treated as evidence of psychological parallelism. In other words, if representatives of different cultures exhibit similar behaviour in a similar situation, they can be assumed to employ similar cognitive processes. This opens up the possibility of discovering cultural variants of universal cognitive processes through comparative studies of behavioural responses to similar or comparable situations (collected, for example, through controlled psychological tests or qualitative methods). Of course,

3 In this regard, the fate of the few anthropological studies of animals (Elizabeth Thomas's 1993 study on dogs is probably the most well-known) is rather instructive. These studies have been heavily criticised by ethologists for anthropomorphism because they interpret animal behaviour in terms of underlying strategies, values, and judgments (e.g. Mitchell et al. 1997: xvi). Anticipating this criticism, Thomas argued, in the introduction to her work, that a certain degree of anthropomorphism is unavoidable if the anthropological approach is to be adopted.

the application of this general epistemological principle is not straightforward. For example, particular attention needs to be paid to ensuring the true comparability of the situations: indeed, these situations should not only look similar to the researcher, but also be perceived similarly (e.g., seen as containing similar or comparable sets of cognitive tasks) by the people under study (Vijver et al. 2011). Still, psychocultural universalism clearly offers the possibility of going beyond behavioural descriptions of the ethological type and, therefore, it is not epistemologically sterile.

Despite the fact that all the theoretical positions that belong to the family of psychocultural universalism share the fundamental thesis described above, they still vary substantially in their conceptions of how deep the culturally induced variation on cognitive universals can be, as well as how it exactly "plays out." A useful way to conceive of the differences between various universalistic positions, and methodological approaches related to them, would be to divide them into four groups broadly following the recent discussion by Hazel Rose Markus and MarYam Hamedani (Markus and Hamedani 2007).[4] The first and the oldest of these groups is referred to as the *personality dimensions group* (the "dimensional approach" in Markus and Hamedani 2007:14–15). The positions that make up this group assume that culture influences cognition through structuring the personality variance among its carriers along one or several personality dimensions. The classic example of this is Ruth Benedict's (1934) belief that cultures differentially shape personalities along Apollonian/Dionysian lines. Other examples of dimensions can be temperament, individualism/collectivism (Hofstede 1980), and independency/interdependency (Kitayama et al. 2007). An important peculiarity of these positions is that neither the postulated personality dimensions themselves nor the tendency of individuals to vary along them are usually considered cultural products. The mechanisms by which culture influences personality, as far as they are discussed at all, consist in creating social settings and structures (norms, values, forms of socialisation, etc.) that foster the development or expression of personality along certain dimensions and inhibit it along others. Therefore, according to these positions, culture does not create cognitive variation, but rather plays on variation that "naturally" exists among humans by choosing "from the great arc of human potentialities," as Ruth Benedict put it (Benedict 1934, quoted in Moore and Sanders 2006:78). In this sense, these positions are probably closest to psychological absolutism. These positions share two further tendencies: to produce broad classifications of cultures along the chosen personality dimensions (to speak about, for example, "Apollonian" and "Dionysian" cultures, or "collectivist" and "individualist" cultures); and to reduce very diverse and broad complexes of culturally specific behaviour and thinking to these dimensions.

---

4 Markus and Hamedani proposed five groups of universalistic approaches. However, in our opinion, two of them – the "cognitive toolkit" and the "ecocultural" approaches – can be combined, since the difference between them is in the details of the causal relations postulated rather than in their general layout.

The second group can be referred to as *sociocultural models* (the "sociocultural models approach" in Markus and Hamedani 2007). The positions that fall under this rubric have been widely accepted and explored over the last decades of the twentieth century, particularly in cognitive anthropology (D'Andrade and Strauss 1992; D'Andrade 1990; Shore 1998). The essential idea behind them is that culture influences cognition by supplying specific models or blueprints of how cognitive processes should be applied or realised and, in doing so, supports the formation of culturally specific forms of behaviour and fosters mutual understanding between the members of the cultural group. In the cases of representation and decision making, the two cognitive processes that have been most researched from these theoretical positions, these models can take a number of forms: categorisation models (the pre-made systems of categories into which the flow of percepts is classified and, therefore, meaningfully represented – see Atran 1990; Ellen 2006; Shore 1998); value systems that inform aims of action (Shore 1998); reasoning and behavioural schemes and scripts (more or less algorithm-like "programs" of reasoning and behaving that are activated for certain cognitive tasks – D'Andrade 1990; Holland and Quinn 1987; Shore 1998; Strauss and Quinn 1998); or rules of thumb (particular kinds of schemes that directly assign cognitive or behavioural "outcomes" to certain stimuli in the way of "if A then B"). Bourdieu's widely known theory of practice (Bourdieu 1977), which is based on the notion of *habitus*, defined as culturally induced "structured structures predisposed to function as structuring structures, that is, as principles of the generation and structuring of practices and representations" (Bourdieu 1977: 72), can also be included in this group. The sociocultural models can be implicit or explicit, but they are in any case "taken for granted …, widely shared by the members of a society …, and play an enormous role in their understanding of the world and their behaviour in it" (Holland and Quinn 1987: 4). In this sense they can be viewed as intersubjective collective representations (Durkheim 2008 [1912]). Note that the sociocultural models theories go much further than the personality dimensions group in acknowledging the formative influence of culture on cognition: in the case of cognitive models, culture indeed creates cognitive variation of its own rather than just structuring the "natural" variation that occurs across individuals. However, this culturally specific variation is assumed to come from structuring cognitive processes that are believed to be universal and uniform. In other words, sociocultural models are perceived as recipes specifying how universal cognitive processes are to be applied to particular cognitive tasks. No culturally induced variation in these processes themselves is assumed here.

The third group of theoretical positions, which can be referred to as the *cognitive toolkit* approach, takes the influence of culture on cognition yet one step further by proposing a culturally induced variation. Although the proponents of this approach still believe that basic cognitive processes are universal and, therefore, stay within the framework of psychocultural universalism, they do allow that culture can modify cognitive processes to a certain extent. As Richard Nisbett explained:

> We may say that even if all cultures possessed essentially the same basic cognitive processes as their tools, the tools of choice for the same problem may habitually be very different. People may differ markedly in their beliefs about whether a problem is one requiring use of a wrench or pliers, in their skill in using the two types of tools, and in the location of particular tools at the top or the bottom of the tool kit. [...] Another way that cognitive processes can differ is that cultures may construct composite cognitive tools out of the basic universal toolkit, thereby performing acts of elaborate cognitive engineering. (Nisbett et al. 2001: 306)

Therefore, while representatives of the cultural models group believe that culture can influence cognition by specifying how universal cognitive "tools" should be applied to solving cognitive tasks, the representatives of the cognitive toolkit group go further, suggesting that culture can also determine which "tools" should be applied, as well as the design of the "tools" themselves. Furthermore, according to Nisbett et al. (2001: 306), to the extent that "the frequencies with which the very most basic cognitive processes are used will differ greatly," the "degree and nature of expertise in the use of particular cognitive processes will differ and tacit or even explicit normative standards for thought will differ across cultures." In other words, culture can actively shape the cognitive toolkit itself. The scholars who share this basic assumption still differ greatly in their opinions on how this shaping takes place and how "deep" it can be. Thus, one widely shared position assumes that the cognitive toolkit is shaped during the course of enculturation and socialisation in a particular ecological and social environment that routinely exerts a set of specific pressures on an individual. By responding to these pressures on a day-to-day basis, the individual applies some cognitive tools more frequently than others and, therefore, develops some universal cognitive processes to a greater extent than others. A specific set of "preferred" or "most developed" tools (or particular variants of such tools) – a culturally specific cognitive toolkit – is produced in this way to allow effective behaviour in this particular ecological and/or social habitat (Berry 1976; Whiting et al. 1979; Witkin and Berry 1975; Witkin and Goodenough 1981).[5] This position differs significantly from that of the cultural models group in the sense that no collective representations or other intersubjective shared entities are involved in the interplay between culture and cognition.

Another position within the cognitive toolkit approach – one that acknowledges the role of shared representations – has been taken by Richard Nisbett and his group. In their opinion, the metaphysical beliefs about the nature of the world and its functioning that are shared among the members of a cultural group can determine what they call "tacit epistemologies" – sustainable patterns of perception and thinking at the level of the individual (Nisbett et al. 2001). These epistemologies can foster reli-

5 Markus and Hamedani find it useful to separate this position from the cognitive toolkit group and identify it as a distinct group (that of "ecocultural approaches" – see Markus and Hamedani 2007: 18–20). We, however, find this additional classification not entirely justified and certainly unnecessary for the purposes of this book.

ance on certain cognitive tools rather than others and, therefore, affect the cognitive toolkits. This position leads these researchers to openly deny the basic distinction between cognitive process and cognitive content: in their opinion, "metaphysics, epistemology [which are usually thought of as the content of cognition – *K.I.* and *M.D.*] and cognitive processes exist in mutually dependent and reinforcing systems of thought" (Nisbett et al. 2001: 306).

The theoretical position employing the notion of "distributed" or "extended" cognition, which has been developed in cross-cultural research primarily by Edwin Hutchins (Hutchins 1995, 2010), can, in our opinion, also be included in the cognitive toolkit group. According to this position, many cognitive tasks that people encounter in the "wild" (e.g., outside laboratory settings) are solved by so-called cognitive systems (not to be confused with systems of processes or domains that make up individual cognition). These systems can exceed the boundaries of individual cognition. In other words, the cognitive process is performed and cognitive tasks are solved not by a single individual, but rather by a system of elements, one of which is the individual. The ability to solve a cognitive task is an emergent property of this system. The other elements can be material objects (for example, special equipment for computing or observing) as well as other individuals (as in Hutchins's example of teamwork during marine navigation). Of course, culture can and does play a role in determining what cognitive systems are available for solving particular cognitive tasks and, therefore, can greatly alter the processes that lead to accomplishing those tasks.

Finally, the fourth group of universalistic theoretical positions is largely derived from the "situated cognition" and "embodied cognition" approaches in cognitive science, and owe much of their theoretical background to James Gibson's ecological theory of cognition (Gibson 1986). These approaches tend to deny the distinction between cognition and action, as well as between mind and body. The main theoretical idea underpinning them is that a significant portion of behaviour manifested by people in everyday situations, including culturally specific behaviour, is not the result of high-level cognitive processes involving categorisation, representation and operations on representations (such as inference). Instead, behaviour is bound to the situation in which it occurs and can be better understood as acting on affordances the actor perceives in a given situation, taking into consideration his or her aims and desires. In other words, the behaviour stems directly from the actor's attention to situational affordances. From the viewpoint of this general approach, the impact of culture upon cognition consists not so much in shaping the cognitive processes (designing the cognitive toolkit) or guiding their application to cognitive tasks, but rather in guiding the perception of situational affordances – the "education of attention," as Tim Ingold (Ingold 2000, 2003) often puts it – as well as in shaping behavioural patterns that represent efficient ways of acting on them (embodied responses). In our opinion, theoretical positions of this kind still fall within the framework of psychocultural universalism, as situated and embodied cognition are believed to be universal mechanisms producing behaviour. Positions representing this group, although different in their details, have been recently taken up by Jean Lave (Lave 1988), Tim Ingold (Ingold 2000) and some others (Nygren 1999; Wenger and Lave 2006; Wenger 1999).

It should be clear that the groups of theoretical positions described above are not incompatible. In truth, the idea that culture can produce specific cognitive toolkits does not exclude the possibility that it can also provide sociocultural models that would guide the application of these toolkits. In addition, neither the former nor the latter would exclude the possibility that culture can also play on "natural" variation in human personality. Although the adherents of situated cognition do insist that much of everyday human behaviour does not result from higher cognitive processes, most of them do not assert that these higher processes do not exist or that they are unimportant to produce action. Therefore, they would perhaps be quite comfortable with the idea that culture can influence these processes in a number of different ways. Furthermore, combining theoretical positions can well lead to their further clarification and development. As previously mentioned, Nisbett and his group were able to employ the notion of collective representations that lead to "shared epistemologies" (which, in our opinion, represent cultural models in a very classical sense) to explain the origin of culturally specific cognitive tools and postulate testable hypotheses about their details. Similarly, the particularities of a specific cognitive toolkit shared by members of a certain cultural group can potentially explain a great deal about their intersubjective sociocultural cognitive models, as well as, most likely, why their culture favours specific personality traits.

Taking all this into account, this book adopts a broad approach of psychocultural universalism without claiming exclusive allegiance to any group of theoretical positions within it. Although each particular study described in this book has been informed by (and, therefore, can be associated with) one of the theoretical positions described above, the general argument we wish to present is essentially a generalisation derived from most, if not all, of the positions described above. Indeed, the main point we wish to demonstrate through these studies is that nomadic reindeer herding as an "economic enterprise" that presupposes a certain way of life and involves specific relations between people, several animal species, and the natural environment, routinely poses a set of rather specific cognitive tasks for its practitioners. In order to solve these tasks effectively on a day-to-day basis, reindeer herders have developed specific cognitive processes (cognitive tools), as well as models to guide their application in particular situations. These processes and models are likely to mutually influence each other and together they constitute an essential part of reindeer herding "culture" (as Goodenough defined it). They can substantially influence the economic and social behaviour of those engaged in reindeer herding, and as far as personality traits are concerned, create important differences between peoples. As a result, adapting and developing these mechanisms (processes and models), is an essential part of socialisation for a reindeer herder.

The general approach we adopt throughout most of this book consists of four basic steps. First of all, we identify certain cognitive tasks that reindeer herders routinely encounter in their everyday practice and describe details of context in which these tasks are encountered. Second, we describe the cognitive processes that, as the studies in cognitive science suggest, can be used to solve these tasks. Third, we try to identify, through observation and experiments, which of these processes are likely to be employed by the herders and which particularities they might have in each case. Finally, we attempt to describe the specific ways in which these processes are applied to the concrete tasks identified.

1.4 Study groups

The studies described in this book have been conducted among nomadic reindeer herders, who belong to two distinct ethnic groups: the Nenets and the Izhma Komi.[6] The Nenets (previously called Samoyeds) number approximately 41,000 people according to the 2002 Russian census.[7] They live in a vast territory in Arctic Eurasia that stretches along the Arctic Ocean from the Kanin Peninsula and the White Sea on the west through to the north of the European part of Russia, the Ural mountains, and north western Siberia up to the Gydan Peninsula and the Yenisei Bay to the east. Throughout this territory, Nenets reindeer herders, who currently account for somewhere between 25 and 50 percent of the entire group, pasture more than 800,000 semi-domesticated reindeer[8] in the tundra and northern taiga.[9] Nenets therefore may well be the largest group of nomadic reindeer pastoralists in the modern world with respect to both the number of herders and the number of reindeer.

According to archaeological and linguistic evidence, the Samoyedic-speaking predecessors of the modern Nenets occupied the current territory of Nenets reindeer herding at least since the end of the first or the beginning of the second millennium AD (Dolgikh 1970) and were already familiar with the domesticated reindeer (Pomishin 1990). Their more than one thousand-year-long history of populating and

---

6 Nenets culture and society has long been an object of anthropological investigation. Of particular importance are the works of Ludmila Khomich (1995) and Andrei Golovnev (1995), which provide an overview of earlier research and discuss in detail most of the topics related to traditional Nenets culture. In English language, books by Igor Krupnik (1993), Andrei Golovnev and Gail Osherenko (1999) and Florian Stammler (2005) provide accounts. Komi reindeer herders have thus far attracted much less attention from ethnographers. To the best of our knowledge, in English language, Joachim Otto Habeck's 2005 published thesis is a rare work focused exclusively on this group. Some aspects of their culture and history are discussed in the classic Russian-language works of Belitser (1958) and Konakov and Kotov (1991).

7 The data are available on-line at http://www.perepis2002.ru/ct/doc/TOM_04_02.xls.

8 This estimate originates from the 2006 Russian Agricultural Census as reported in Skolkin 2008.

9 *Taiga* refers to subarctic forest dominated by coniferous trees.

exploiting the harsh arctic environment has resulted in the development of a unique complex of material culture and technological knowledge – the so-called Samoyedic reindeer herding complex – which has proven highly efficient in supporting the nomadic way of life based on exploiting the resources of reindeer (Khomich 1995; Vasilevich and Levin 1951; Yuzhakov and Mukhachev 2001). This complex, which was adopted completely or partly by many other reindeer herding groups (Startsev 1926; Zherebtsov 1982), includes such elements as the conical nomadic tent (*chum*) covered by two layers of reindeer skins; several types of wooden reindeer sledges used throughout the year for transporting people and their belongings; male and female coats (*malitsa* and *panitsa* respectively) made of reindeer fur and highly adapted to the arctic climate; basic techniques of catching reindeer with a leather lasso, harnessing them to sledges in a fantail-like manner, and steering them with a wooden stick (*khorey*); basic techniques of building temporary corrals for reindeer out of sledges and rope nets; and the use of herd dogs to control herds of reindeer on pastureland (Khomich 1995; Yuzhakov and Mukhachev 2001). Nenets have also bred their own breeds of domesticated reindeer (the so-called Samoyed reindeer) and herd dogs (Samoyed husky) (Pomishin 1981).

Taking into account this rich reindeer herding heritage of the Nenets, it is ever more surprising that, up until the beginning of the eighteenth century, their economy was based, in fact, on mobile migratory hunting rather than on reindeer herding (Krupnik 1976, 1993). The few domesticated reindeer, not more than a couple dozen per household (Krupnik 1976), were used predominantly for transportation during seasonal migrations and hunting expeditions, and were normally never milked or slaughtered for food or skins. All subsistence products as well as goods for exchange and the payment of taxes were obtained from hunting (Krupnik 1993). The emergence of the truly nomadic pastoralist economy among the Nenets took place relatively recently – not more than 300 years ago – as a result of a rather complex process of cultural and economic transformation (Krupnik 1976). In this process, the second group of reindeer herding pastoralists on which the studies presented in this book are focused – the northern (Izhma) group of Komi – most likely played an important rôle.

Compared to the Nenets, the Komi are newcomers to the Eurasian Arctic. The predecessors of the modern Komi reindeer herders came to the Arctic from the south no earlier than the sixteenth century AD (Konakov and Kotov 1991; Lashuk 1958). The historical homeland of the Komi people – the basin of Vychegda River in the northeastern part of European Russia – is situated in the southern taiga, and most Komi rermain in this region. Archeological sources indicate that the predecessors of the modern Komi in this region were relatively sedentary and their economy was based in approximately equal proportions on slash-and-burn agriculture and hunting (Savelieva 1971) that, since about the fifteenth century AD, took the form of commercial trapping of fur-bearing animals (Konakov and Kotov 1991; Konakov 1983). The migration of a part of the Komi people northward was related, most probably, to a decrease in the population of fur-bearing animals in the southern taiga caused by, among other things, excessive trapping (Konakov and Kotov 1991). Until the beginning of the eighteenth century, their residence in the Arctic was limited to the northern taiga of northeastern Europe (the basin of the Pechora River), where they established several permanent settlements. The oldest and largest of these settlements is the town of Izhma, from which the migrant group of Komi has taken its name – Izhma Komi (Lashuk 1958). As Russian sources of that period clearly indicate, the migrants failed to develop agriculture as an important branch of their economy in their new homeland (Lashuk 1958), probably because the efficiency of the agricultural techniques they brought from their region of origin was limited in the Arctic (Konakov and Kotov 1991). However, they maintained their settled way of life, which was supported predominantly by commercial trapping (Konakov and Kotov 1991; Konakov 1983).

The transition of both the Nenets and the Izhma Komi to full-scale nomadic reindeer pastoralism – understood as a kind of productive economy based on raising reindeer predominantly for meat and characterised by specialized herding techniques and patterns of regular migrations aimed at maximizing reindeer productivity – occurred during the eighteenth century (Istomin 2004; Krupnik 1976, 1993). Fortunately, official Russian documents reflect this transition relatively well (Istomin 2005; Krupnik 1976). It appears that this new economy emerged in northeastern Europe simultaneously among the Komi and the Nenets, and rapidly spread to the Nenets living to the east of the Ural mountains. At the same time, the Izhma Komi adopted the nomadic way of life and fanned out into the tundra, first on the western side of the Urals and later moving over to the eastern side. The reasons for this explosive transition to the new economic model are still debated (Golovnev and Osherenko 1999; Golovnev 1995; Krupnik 1976, 1993; Pomishin 1990). In a previous work, Istomin (2004) argued that it was caused by a complex of ecological, economic, and social factors (Istomin 2004). Thus, the adoption of demi-domesticated reindeer by the Komi was most likely triggered by the depletion of the population of fur-bearing animals in the proximity of permanent Komi settlements, which necessitated an increase in the amount of territory needed for hunting expeditions. The adoption of transport reindeer herding from the Nenets offered a fairly effective solution to this problem and allowed Komi trappers to include new areas of the northern taiga and southern tundra into their hunting territories. This, however, led to competition and conflicts, sometimes rather violent, between the newcomer Komi and Nenets who had been exploiting these territories for centuries. These conflicts, reported in increasing numbers in Russian documents over the first half of the eighteenth century (Istomin 2004, 2005), caused a significant redistribution of Nenets hunting groups and their amalgamation together in order to provide more effective resistance. The growth of the individual herds among the Nenets, which is also reported in documents from this period (Istomin 2004; Krupnik 1976), may also have been caused by this change in the sociopolitical situation. Furthermore, the growth probably reflected an increased need for transportation, as well as an insurance strategy at a time when animals were often captured and stolen by the enemy.[10] The ecological pressure caused by increased hunting and grazing of domesticated reindeer led to the depletion of the wild reindeer populations, the main object of hunting in the tundra, and to its eventual disappearance in the beginning of the nineteenth century (Baskin 2009). This apparently necessitated the start of regular slaughters of semi-domesticated reindeer for food and skins (previously practiced by the Nenets only in times of famine) among both Nenets and Komi. In doing so, the transition to nomadic pastoralism was, therefore, completed.

This account of the origin of nomadic reindeer herding among the study groups represents just one possible scenario that is consistent with available historical evidence. One fact, however, is indisputable: nomadic reindeer herding among both the Komi and Nenets clearly has its roots in the Nenets transport reindeer herding of previous centuries and is based on the Samoyedic reindeer herding complex: both Komi and Nenets reindeer herders construct a similar type of nomadic tent (*chum*), use the same types of reindeer sledges and reindeer harnesses, and wear the same type of winter clothes made of reindeer fur. Furthermore, the names for all these items, as well as the age and sex classification of domesticated reindeer (Startsev 1926), in both the Komi and Nenets languages, are the same and are obviously of Nenets origin.

Despite the common roots of reindeer herding nomadism among the Nenets and Izhma Komi, the more recent development of this type of economy has taken quite different paths among the two groups. As the Komi were involved in a market economy long before becoming reindeer herding nomads, and, as we have just argued, the new economy emerged to replace commercial trapping, their reindeer pastoralism immediately took on the form of a highly commercialised activity (Istomin 2004; Kertselli 1911; Krupnik 1976). The animals were raised primarily to sell their products (meat, skins often processed into suede, blood, etc.) on the market for commercial profit (Konakov and Kotov 1991). Furthermore, the regular and well-established contacts with southern traders allowed many Izhma Komi herders to become mediators between the Russian market and the nomadic population of the tundra (primarily the Nenets) that was less involved in market relations (Kertselli 1911; Zhuravski 1909). Because the dependency of the tundra nomads on goods imported from the south (e.g., metal tools, firearms and ammunition, food staples such as flour, etc.) constantly increased through the nineteenth century (Khomich 1995), this position secured for the Komi tremendous profit as well as economic and political power. Izhma Komi herders thus quickly acquired an economically and politically dominant position, particularly in the tundra areas of northeastern Europe, leading to the rapid, large-scale expansion of their herds and of the territory over which they performed migrations. By the second half of the nineteenth century, Izhma Komi herders could be

10 Stories of the Komi capturing or stealing Nenets reindeer abound in the accounts of travelers who visited the region at the end of the eighteenth and beginning of the nineteenth centuries (Islavin 1847; Schrenck 1848).

found throughout the entire Arctic region of northeastern Europe and western Siberia, from the Kanin peninsula in the west to the mouth of the river Pur in the east (Konakov and Kotov 1991). This area of migration coincided almost exactly with that of Nenets, and the migration routes of Komi and Nenets reindeer herders overlapped. Furthermore, by the end of the nineteenth century, Izhma Komi reindeer herders had expanded their territory even further to the west by colonising the Kola Peninsula (Konakov and Kotov 1991). The commercial character of reindeer herding pastoralism among the Komi has had an important impact on their reindeer husbandry techniques, particularly migration and grazing patterns, which became particularly well-adapted to managing very large herds of reindeer and substantially differ from those of the Nenets. These differences will be discussed in detail in the following chapter. The social organisation of reindeer pastoralism among the Komi was also affected by commercialisation: starting in the second half of the nineteenth century, Izhma Komi reindeer herders became increasingly dependent on hired labor to manage the grazing of their large herds. Sergei Kertselli, one of the most renowned Russian experts on the reindeer herding of the early twentieth century, aptly summarised these particularities of reindeer herding nomadism among the Komi under the rubric "tundra capitalism" (Kertselli 1911).

In contrast to the Komi, nomadic reindeer herding among the Nenets continued to be the basis of the "domestic mode of production." Its products were consumed mostly within the household, while the small portion that found its way to the market was for the most part exchanged for "southern" goods that were needed for domestic consumption. For these reasons, the number of reindeer owned by Nenets herders, after a brief period of growth in the eighteenth century, remained constant throughout the nineteenth century (Kertselli 1911). In many households, particularly those in western Siberia, reindeer herding was combined with subsistence hunting and fishing, with the role of these economic activities being approximately equal. The need to combine herding with hunting and fishing affected herding techniques and migration patterns in several ways, as will be discussed in the next chapter.

The relative advantages and disadvantages of the nomadic economies of the two groups were turned upside down after the communist revolution in Russia in 1917. The economic crisis in Russia caused by the revolution and the civil war that followed had a tremendous negative impact on the commercial reindeer herding of the Komi. Furthermore, the reliance of many Komi herding households on hired labor made them excellent targets for the new communist government's campaign against "exploiters." The collectivisation of reindeer herding in the beginning of the 1930s, during which private reindeer herds were confiscated and given over to newly organised collective enterprises (*kolkhozes*) and, later, state-owned enterprises (*sovkhozes*) that employed reindeer herders as hired workers, represented the final and fatal blow to "tundra capitalism" (Istomin 2010). It seems as if the impossibility of practicing reindeer herding as a private commercial enterprise made it a pointless activity for many Komi. Many wealthy reindeer herders preferred to slaughter their herds rather than have them confiscated (Istomin 2010). Many others decided to settle and take poorly paid jobs in the villages and towns rather than continue as hired herders working for *sovkhozes*. As a result, by the end of the 1950s, the number of active Komi reindeer herders in northeastern Europe (the core area of Komi reindeer herding) was approximately half of what it had been at the beginning of the twentieth century, while the number of reindeer declined to about one-third (Istomin 2010). The numbers of both reindeer and herders continued to decrease, although less dramatically, throughout the Soviet period – the number of reindeer stabilised at the beginning of the 1970s before falling again in the late 1980s (Istomin 2010).[11]

Although collectivisation and the associated economic disturbances also had a negative impact on the Nenets herders, it was much less pronounced. Following collectivisation their herds decreased by approximately 30 percent (Slezkine 1996) before starting to grow again in the 1960s. More importantly, the exodus of people from reindeer herding was not significant; in fact, the overall number of herders actually increased. It can be argued that one important reason for this was that, in contrast to the Komi, the Nenets have always viewed reindeer herding nomadism more as a way of life than as a commercial enterprise. The reaction of the two groups to the collapse of the Soviet Union and the socialist economic order is an excellent illustration of this fact. Once the possibilities of private ownership and enterprise were opened again, many nomadic households among the Nenets used the opportunity to leave the state-owned and managed reindeer herding enterprises and re-establish themselves as subsistence reindeer herders managing private herds. As a result, the number of reindeer pastured by Nenets in western Siberia grew significantly (in some regions it doubled) between 1991 and 2000 (Jernsletten and Klokov 2002). Judging from the results of the 2008 Russian Agricultural Census, it appears that this trend continued over the first decade of the twenty-first century. Significantly, the absolute majority of reindeer pastured by the Nenets now represent private property (Jernsletten and Klokov 2002). No comparable process was evident among the Komi. While there has been a small increase in the number of privately owned reindeer since 1991, this has

11 Precise data on the change in numbers of Komi herders and their reindeer over time are very difficult to obtain because the available data in most regions populated by Komi do not differentiate between Komi herders and herders from other groups (e.g., Nenets). More or less reliable numbers could be obtained only for the Bol'shezemel'skaya tundra, the core area of Komi reindeer herding, where data on Komi and Nenets were kept separately. These data suggest that in the beginning of 1920s (that is, before collectivisation) there were as many as 4,500 active Komi reindeer herders with about 600,000 reindeer in this region alone. By the end of the 1950s, the number of Komi herders in the region had dropped to 2,300, while the number of reindeer they herded decreased to 220,000. By the end of the 1990s, Komi reindeer herding enterprises in the region employed 1,200 herders who herded 160,000 reindeer. Information from the 2008 Russian Agricultural Census allows us to estimate the total number of active Komi reindeer herders in Russia at approximately 3,000, and they herd approximately 250,000 reindeer.

been accompanied by an even more significant decrease in the number of reindeer belonging to the state-owned enterprises, resulting in an overall decline in the total number of reindeer pastured by Komi when compared to the late Soviet period (2008 Agricultural Census). What is appears even more imteresting to note, is no significant outflow of herders from state-owned enterprises to private herding was evident. Furthermore, the process of abandoning reindeer herding, which had slowed down during the 1990s due to the economic crisis in Russia and the absence of jobs in the settlements, has evidently resumed during the beginning of the economic recovery in the first decade of the twenty-first century. About 3,000 active reindeer herders with a total of approximately 250,000 reindeer, most of which belonging to reindeer herding enterprises (2008 Russian Agricultural Census), represent all that is left of the "tundra capitalists" who once dominated the entire western part of the Eurasian Arctic.

As it is clear from the historical account above, only a minority (in the case of Komi, a small minority) of the representatives of both ethnic groups currently live the life of reindeer herding nomads. Furthermore, even those representatives of the two ethnic groups who are reindeer herders can differ in the details of their way of life and herding technology from one region to another. For example, migration patterns and human-reindeer relations among Nenets reindeer herders of the Yamal peninsula seem to be closer (although not entirely similar) to those described below for Komi reindeer herders of Bol'shezemel'skaya Tundra then to those of Taz Nenets described in this book (Florian Stammler and Elena Liarskaya, pers. communication). Similarly, the way of life and the kind of animal husbandry practised by the group of Komi herders currently living on the Kola Peninsula has similarities with those of Scandinavian Saami: these Komi herders do not migrate with their reindeer in summer and allow the animals to freely graze on their own; in winter they regularly check (from once per week to once per month) and round up their herds, but do not move with them (Istomin 2017). Finally, although the majority of reindeer herders from both ethnic groups still practise their languages, some nomadic groups have switched to using the Russian language in their day-to-day life.

It is important, therefore, to explain whom we refer to as Komi and Nenets in this book and what we mean by Komi and Nenets culture. Unless it is explicitly stated otherwise, the term "Komi" refers to Izhma Komi nomadic reindeer herders migrating in the eastern part of Bol'shezemel'skaya tundra, that is in the sites in which we carried out our fieldwork. Similarly, the term "Nenets" refers to Nenets reindeer herding nomads living in the Taz tundra, the other field sites in which we worked. A more detailed description of our field sites will be given shortly. Before proceeding to it, however, it is important to stress that such an understanding excludes treating the concept of culture as "ethnic culture", something a person possesses by virtue of his/her being born and raised in a particular ethnic group or having a particular ethnic background. This is important to stress, because such an understanding of the concept is popular in Russia and certain European schools of anthropology. Our use of the concept, however, is more concrete and corresponds more closely to our way of using the group names: in our reading, culture is a set of features a person possesses by being born and socialised in a particular community with a certain way of life, certain economy (in our case reindeer herding), a certain social organisation, a certain language, etc. In other words, the phenomena we refer to in our book as the culture of Nenets/Komi reindeer herders represents something Nenets/Komi reindeer herders possess not because they happen to have a Nenets or Komi ethnic background (despite their having one), but because they are reindeer herders who have a particular reindeer herding technology, a particular way of life and, to some extent, a particular language.

1.5 Field sites

The studies among the two reindeer herding groups that will be presented in the following chapters were conducted in two separate field sites. The research among the Nenets reindeer herders was carried out in the Taz tundra, situated in the southern part of the Gydan peninsula in western Siberia. This region, known as *Tasu' yawa'* in the Nenets language, administratively represents the Tazovskii and Gaz-Salinskii *selsovets* (rural administrative units) of the Tazovskii *raion* (district) in the Yamal-Nenets Autonomous *Okrug* (province) of the Russian Federation. The whole region is situated approximately 100 kilometers north of the Arctic Circle, in a transitional area between the taiga zone and the tundra zone. The river Taz (*Tasu-yaha* in Nenets), one of the largest rivers in Siberia, is the most pronounced natural feature of the region. It crosses the region from east to west and then flows northwest to the Arctic Ocean, and constitutes a huge labyrinth of interconnected channels and lakes separated by numerous islands of different sizes. All these channels and islands together form an enormous floodplain that ranges from forty to seventy kilometers wide. The numerous channels forming the main river, as well as some of its tributaries, are the natural spawning grounds for several valuable species of fish, including Siberian sturgeon, Siberian white salmon, and Siberian whitefish. The rest of the region outside the Taz river floodplain consists of flat tundra covered mostly by lichens and dwarf birches.

Although the region was claimed by the Russian crown in the seventeenth century and was visited (sometimes rather irregularly) by Russian tax-collecting and trading expeditions thereafter (Miller 1941), it was only in the early years of the twentieth century that the first permanent Russian settlement in the region – which is now the town of Tazovskii – was established close to the mouth of the Taz river (Lehtisalo 1956). The major influx of non-aboriginal settlers into the region dates back to the 1960s, when large amounts of natural gas were discovered. At present there are three permanent settlements in the region: besides the town of Tazovskii (*Halmer Sede Harad* in Nenets), there are the villages of Gaz-Sale and Nakhodka, which serve as home bases for gas drilling explorations. In addition, there is a small village, Tebei-Sale, the former *baza osedlosti* (sedentarisation base) of the former Tebeisalinskii reindeer-herding *kolkhoz*. The *kolkhoz* was reorganised into the Tazovskii *sovkhoz* in the 1970s, and the village was officially abolished. Still, there are five permanently inhabited houses there. Other locations with permanent populations include three gas drilling stations and a number of gas drilling towers distributed throughout the tundra. The villages and the drilling stations are connected to each other by seasonal roads, which can be used from December through April when frozen. The construction of the first permanent road in the region, which connected the town of Tazovskii, the village of Gaz-Sale and the nearest railway station situated about 250 kilometers to the south, was completed in 2007. The town of Tazovskii, with a population of 3,000 people, is especially important as a trade center where the aboriginal population of the tundra comes into contact with the sedentary world. The central office of the reindeer herding collective farm, the boarding school for children of nomadic families and the local representatives of the provincial departments of aboriginal affairs are also situated in Tazovskii.

The nomadic population of the region consists exclusively of Nenets and is divided into two groups: reindeer herders (*ty perta*, pl. *ty perta'*) and fishermen (*hala iorta*, pl. *hala iorta'*). It should be stressed that this division is emic in the sense that it represents the way the Nenets see themselves. In fact, for an outside observer this distinction is not very clear. Many fishermen also have considerable herds of private reindeer and can, therefore, also be regarded as reindeer herders. On the other hand, all people who refer to themselves as herders also fish. What this division actually reflects is the two distinct modes of land use. Those *hala iorta'* who have reindeer usually live as nomadic reindeer herders from December to May. However, at the end of May, when the reindeer calving period has ended, they usually hand their reindeer over to the *ty perta'* and erect their chums in the fishing grounds (*mara*, pl. *mara'*) situated in the Taz river floodplain. Normally every family (*chum*) occupies its own *mara*. They fish there until the middle of July, which is when the spring spawning of fish in the river Taz ends. In the middle of July most fishermen dismantle their chums and move in boats down the Taz river to the sea coast, where the summer sea fishing grounds – the *ier mara'* (main *maras*) – are situated. Near the sea, a number of fishing families pool their workload and live together, thereby creating "villages" that can consist of more than ten chums. The sea fishing continues until the end of August or the beginning of September, when the *hala iorta'* start to make their way back to the upper parts of the Taz in order to benefit from the autumn spawning of fish in the river. The autumn fishing is very important because most of the fish will be given to the *ty perta'* as payment for taking care of the reindeer during the summer. This payment usually takes place in November. From then until the second half of May the families of fishermen move with their herds in the area directly adjacent to their *mara'*.

In contrast to the *hala iorta'*, most of whom live in isolated households except when they are at their summer fishing places, the *ty perta'* usually migrate in groups referred to as "camps" (*stoibishe* or *okolodok*) by the Russian administrators. Each camp consists of between four and seven chums. Five of these camps still represent the *brigadas* (herding units) of the former Tazovskii reindeer herding *sovkhoz*. There are also at least five camps of private herders. Membership in a camp is flexible and there is a constant flow of chums between the camps. Each camp of Taz Nenets reindeer herders migrates in a more or less circular fashion within a territory approximately sixty kilometers in diameter, where it pastures its reindeer throughout the year. These territories are not fixed, however, and they are reported to change quite often. Further details on their pattern of migration and the herding technology will be given in the chapters to follow.

The research on the Izhma Komi reindeer herders was conducted in a second study site that included the eastern part of the Bol'shezemel'skaya tundra in the north-eastern part of European Russia. This region, which historically represented a part of the core area of Izhma Komi reindeer herding, administratively belongs to the Intinskii *raion* of the Komi Republic and to the eastern part of the Nenets Autonomous *Okrug*. As Izhma Komi, in contrast to the Nenets of the Taz tundra, have linear rather than circular migration routes, the territories of their migration are long, narrow strips of land (called *olenegonnie koridori* – "herding corridors" – by Russian administrators) that stretch some 200 to 400 kilometers inward from the northern taiga, where the winter pasturelands are usually situated, through the tundra zone all the way to the midsummer pastures on the coast of the Arctic Ocean. Typical migration routes ranging from 500 to over 900km in length – see Dwyer & Istomin (2009) for further details on migration matters and how these changed as result of the rise and fall of Soviet amalgamation and collectivisation. The Izhma Komi herders who provided the data for this book belong to the Bol'shaya Inta reindeer herding enterprise (former *sovkhoz*).[12] They migrate within the herding corridor that was assigned to this enterprise in the 1960s, when it was established through the amalgamation of several smaller collective reindeer herding enterprises (*kolkhozes*) that were created by the collectivisation policy of the 1930s. The winter pastures that make up the southern part of this corridor are situated to the north of the Usa river and to the east of one of its largest tributaries – the Adzva river. The area is representative of the classic forest-tundra zone, with large islands of forest separated by large and medium-sized treeless tundra-like areas. Three "traditional" settlements – the villages of Adzva-vom, Rogovaya, and Petrun – which were established during the period of the peak of Komi reindeer herding in the nineteenth century and once represented important centers of the herders' suede production and trade, are the most important landmarks in this part of the region. Most of the herders have relatives living permanently in one of these settlements, associate themselves with them and live there for more or less extended periods of time during the winter. A seasonal road connects these settlements to each

12 In 2012, Bol'shaya Inta was amalgamated with a few other agricultural enterprises not involved in reindeer herding to produce a new enterprise, the Agricultural Complex "Inta Pripolarnaia".

other as well as to the "modern" city of Inta, which is situated about sixty kilometers to the south across the Usa river. This city, which was established in the 1940s as a center of coal mining and currently has a population of approximately 3000, is the place where the herding enterprise's main office is situated. Starting from this region of winter pastures, the migration paths (*vörga*, pl. *vörgaias*) of the local Komi reindeer herders proceed in a northeasterly direction across the whole Bol'shezemel'skaya tundra to the midsummer pasturelands on the Kara Sea coast. These pasturelands are situated on the Iugorskii Shar peninsula between the military settlements of Amderma and Ust-Kara, each approximately thirty kilometers away.

When we started our work with the Komi reindeer herders began in 2001, the Bol'shaya Inta reindeer herding enterprise included five *brigadas*, each of which consisted of eight to twelve male reindeer herders and two to four females, who were called *chumrabotnitsi* ("*chum* workers") in the documentation of the enterprise. Each *brigada* consists predominantly of members of two core families that are sometimes related to each other by affinal ties. Still, most of the *brigadas* also include one to three male herders who are unrelated to the core families are affiliated with the herding unit by the decision of the enterprise. All the members of a *brigada* live together in one single nomadic conical-shaped (*chum*), which is divided into two halves, one for each of the two core families (see Habeck 2005; Istomin 2000 for details). The practice of two herding families living together and making up a herding unit is quite traditional among the Izhma Komi and dates back to the nineteenth century, when the two families were usually those of the master of the herd and his hired worker (Istomin 2000). Each *brigada* is responsible for a herd of between 2,500 and 3,500 reindeer, most which belong to the enterprise. The private reindeer of the herders are pastured together with those belonging to the enterprise and are differentiated from the latter by special earmarks. The herders usually leave the winter pasturelands in the southern part of their corridor in the beginning of April in order to reach the treeless southern part of the tundra by the beginning of May. After a short stay for the purpose of reindeer calving, which occurs mostly during the first two weeks of May, the herders continue north and usually reach their midsummer pasturelands near the Arctic Ocean by the second half of July. In the middle of August, after about three weeks on the midsummer pasturelands, the herders start their southward movement and usually return to the winter pasturelands by the second half of November. A more detailed discussion of their migration pattern will be given in the next chapter.

1.6 Structure of the book

The chapters that follow communicate the results of our studies among the Nenets and the Izhma Komi reindeer herders between 2001 and 2011. Some of the material included here has been previously published as journal articles or chapters in edited volumes, and some has been presented as conference papers. All of these texts have been substantially revised and expanded in order to construct a more consistent and overarching argument that addresses the object and aims of this book as described above.

This argument starts with a detailed description of how the interaction between the herders, the animals, and the landscape is organised among the two groups (chapter 2). This interaction, in our opinion, makes up the essence of any nomadic pastoralist economy, while the knowledge and skills involved in it represent the core of herding technology that any herder has to internalise and cognitively adapt to. On the basis of this description, we will try to demonstrate what specific cognitive tasks the reindeer herders from the two groups routinely face in their everyday lives. This provides the basis for the developing and testing of hypotheses about the cognitive mechanisms involved in solving these tasks that constitute the contents of the remaining chapters. As has been stated, these mechanisms, as far as they are covered in this book, belong, for the most part, to two cognitive domains: spatial cognition, and deciphering and predicting reindeer behaviour.

Chapters 3 to 5 analyse the mechanisms related to spatial cognition. The first of these chapters reviews the theoretical models accounting for spatial cognition in humans as they have been proposed in geography, anthropology, and cognitive science, and investigates how the empirical material collected among the two groups of reindeer herders is consistent with these models. Chapter 4 focuses on the process and methods of wayfinding and the cognitive mechanisms that likely support them among reindeer herders, and compares them to some other groups of people in the tundra who depend heavily on wayfinding abilities: amateur hunters and helicopter pilots. Special attention is paid to wayfinding errors and the conditions in which they occur among these groups. Further insights into the spatial cognition of reindeer herding nomads are offered on the basis of this study. Finally, chapter 5 adopts a more classical cognitive-anthropological approach to investigate the conceptualisation of landscape among reindeer herders. This investigation provides further details on how the spatial cognition of reindeer herders is related to and used during the interaction between the herders and their animals.

Chapters 6 and 7 take a closer look at reindeer behaviour and the herders' skills in deciphering and predicting it. The focus of the first of these chapters is on the relationship between specific patterns of reindeer behaviour and the ways of organising the interaction between the animals and the herders that are characteristic of a particular reindeer herding system. The main message we would like to get across in this discussion is that the reindeer behaviour that the reindeer herders perceive and have to make sense of is, to a very substantial degree, the product of the specific reindeer herding system of which both reindeer and the herders are a part. This fact, in our opinion, has important implications for understanding the cognitive mechanisms the herders use to decipher and predict this behaviour, as well as their limitations. These mechanisms are investigated in chapter 7, where we assert that they are likely to be an extension of the Theory of Mind, the basic mechanism humans use in the domain of social cognition. This suggestion, if it is correct and holds true for pastoral nomads outside the Arctic, has several important implications for both psychology and cognitive research on the Theory of Mind and anthropological research on those elements of nomadic-pastoralist cultures that are not directly related to their economic activities per se, such as the study of religion and folklore.

The final chapter of the book summarises the arguments made in the preceding chapters and outlines theoretical conclusions that can be drown from them.

2 REINDEER HERDING: INTERACTION BETWEEN PEOPLE, ANIMALS, AND LAND[13]

2.1 Pastoral nomads, animals, and resources: Some existing perspectives

The first step to understanding what it takes one to know and how it takes one to think in order to be a reindeer herding nomad is to clarify the specific cognitive tasks one routinely has to solve to perform efficiently as a reindeer herder. Clarifying these tasks enables a more focused search for the cognitive mechanisms herders use to solve them. It is rather obvious that the nature of these tasks is directly related to the particularities of reindeer herding, and nomadic pastoralism in general, as a specific type of economy and a specific way of life. Therefore, a good way to start identifying these tasks is to review the existing models of and approaches to nomadic-pastoralist systems, and to try to establish what kind of cognitive operations pastoral nomads can be expected to routinely perform on the basis of these models.

In the popular imagination, pastoral nomads are usually depicted as people raising herds of animals for subsistence, and as constantly moving with these herds from one pastureland to another in search of forage and water for their livestock. This depiction, explicitly put forth in the sixteenth century by Rashid Al-Din, dominated scientific studies of nomads in both geography (e.g., Forde 1934; Semple 1933) and anthropology (e.g., Evans-Pritchard 1940; Myres 1941) up until the late 1960s. Despite anthropologists' long engagement with the concept of nomadic pastoralism, however, it was not until the beginning of the 1950s that they became interested in constructing more detailed models of nomadic pastoralist systems based on this general picture (Dyson-Hudson 1972: 6, see also Dyson-Hudson and Dyson-Hudson 1980; McCabe 1994). By the 1960s two different approaches to modeling nomadic pastoralist systems had already crystallised (see Paine 1972: 77). The first one postulated the existence of homeostatic relations between nomadic pastoralists and their environment[14] and viewed nomadic mobility as a basic mechanism for maintaining this homeostasis by leveling the fluctuations (seasonal, etc.) in the supply of resources (mainly forage and water) in any particular place (e.g., Fredrik Barth 1961). The second approach – attributed by Robert Paine (1972: 77) to E. E. Evans-Pritchard (1940) and considered

13 Parts of this chapter have been published in the journals *Human Ecology* (Dwyer and Istomin 2008) and *Nomadic Peoples* (Dwyer and Istomin 2006).

14 It seems that the notion of homeostasis was adopted from the cultural ecology school, which was founded by Julian Steward (1955) and gained great popularity among anthropologists at that time (see Orlove 1980 for a review).

by Paine to be more appropriate for explaining the behaviour of pastoral nomads – denied the homeostatic relationship between pastoralists and their environment. Instead, pastoral nomadism was seen as a strategy adopted by pastoralists to maximise the supply of resources (which could vary rather significantly from season to season and from place to place) for their herds at any particular moment in time.

However, it also became evident in the 1950s and 1960s that resources necessary for the maintenance of livestock (forage and water) were not the only factor nomadic pastoralists had to deal with in their day-to-day practice; empirical studies demonstrated that there were also a number of other biological factors (the presence of tsetse flies (Stenning 1959) is one of the best examples), as well as non-biological factors, including economic and social relations with other groups (nomadic and settled), national governments, and social relations between individuals within a nomadic group (Barth 1960, 1961; Bates 1972, 1973; Lattimore 1940, to mention just a few). Although resource factors were still generally regarded as the main consideration until the end of the 1960s (McCabe 1994: 72), the urge to incorporate additional factors into explanatory models of nomadic pastoralism increasingly came to the fore. One of the ways this was achieved was by adopting a "possibilistic" approach, which was particularly useful for explaining nomadic-pastoralist movement. For example, according to P.H. Gulliver's "Nomadic movements: Causes and Implications" (1975), pastoral nomads obviously moved in response to ecological (resource) factors. However, the manner in which they did so left open many "opportunities of roughly equal pastoral advantage" (Gulliver 1975: 372). In other words, pastoralists had "a degree of freedom of action" through which socio-cultural considerations could (given that ecological conditions were more or less equivalent) determine their course of movement (Gulliver 1975: 372–373). An essentially similar approach was adopted by Neville Dyson-Hudson, who believed that the simplest model of a pastoral system would be that of balancing the response of "coincident populations (human/livestock) … to a natural environment in which grazing, water and disease vectors are the prominent features" (Dyson-Hudson 1972: 19). This model, however, could provide only a rough approximation of the dynamics involved in nomadic-pastoral systems: more complicated models would need to be developed to acquire a more precise understanding of nomadic pastoralism. The first step toward achieving this was to incorporate "two or more sets of coincident populations, which to some extent compete with or at least respond to the presence of each other" (Dyson-Hudson 1972: 19). The second step was to include factors produced by or relative to "other populations exploiting the environment in quite different ways" (Dyson-Hudson 1972: 19). From the viewpoint of this approach, ecological and non-ecological factors affected the behaviour of pastoral nomads at different scales of precision: ecological factors determined the general features of a nomadic pastoralist system (such as seasonal migrations from one ecological zone to another), while non-ecological factors determined the concrete details.

The approach outlined above attracted severe criticism in the 1970s (e.g. Bates 1973; Burnham 1975; Elam 1979; Irons 1974). By that time, ecological studies had convincingly demonstrated that the relationship between nomadic pastoralists and their environment could generally not be viewed as homeostatic, a conclusion that undermined the models of Barth (1961) and Spooner (1972), which had been based on the premise of homeostatis. Some of these anthropological studies in fact demonstrated that in certain nomadic-pastoralist systems – for example, those of the Chomur group of Yomut Turkmen (Irons 1974) and the Hima of Uganda (Elam 1979) – there were no ecological reasons for nomadic migrations. These nomads moved exclusively due to political, economic or social reasons, which suggested that both ecological and non-ecological factors could be of equal importance in shaping nomadic-pastoralist systems, and that any given concrete system could be shaped by all or any of them (Dyson-Hudson and Dyson-Hudson 1980). Consequently, considering that some nomads migrated to raise herds while others raised herds to migrate, anthropologists began to doubt whether the development of a general model of nomadic pastoralism could be achieved, or indeed whether the very concept of nomadic pastoralism could even be maintained (Dyson-Hudson and Dyson-Hudson 1980).

In the second half of the 1980s a new stage of anthropological research on nomadic pastoralism took place. This stage can be best exemplified by the well-known Turkana research project (see Little and Leslie 1999). In contrast to earlier research, this new stage was marked by applying what Orlove (1980) called a "processual approach" using "actor-based models," as opposed to the pattern-oriented "neofunctionalist approach" of the previous research period. Generally, this meant a change of research agenda "from social structure to social process … and from normative and jural aspects to behavioural aspects of social relations," focusing on the factors and processes involved in shaping the individual behaviour of an actor rather than on explaining abstract patterns, such as those of migration (Orlove 1980: 246). As a result, explanatory models of nomadic pastoralism started trying to explain the decision-making processes of nomadic pastoralists and, therefore, came quite close to the models of cognitive processes of the sort focused on in this book. Two approaches in model building immediately crystallised. The first, which Orlove classifies as "microeconomic models" (1980: 246–247), focused on the factors influencing pastoralists' decisions, which left the actual process of decision making, including its cognitive aspects, outside the scope of its analysis. For example, the early anthropological publications of the Turkana Project (e.g. Dyson-Hudson and McCabe 1985; McCabe 1994) calculated the relative numbers of pastoral nomads' decisions made in response to ecological, economic, social, and political factors in order to understand their relative influence on the functioning of the nomadic-pastoralist system. Many other studies of this group (e.g., Adriansen and Nielsen 2002; De Boer and Prins 1989; Coppolillo 2000; Robbins 1998; Schareika 2001; Sieff 1997; Turner 1999) belonged to different branches of the human environmental ecology school (see Cronk 1991; Cronk et al. 2000). Some of them used models that had been developed in animal ecology, such as the optimal foraging model (e.g., De Boer and Prins 1989), whereas others (e.g., Sieff 1997) used dynamic optimality models to explain and predict nomadic pastoralists' behaviour as maximising certain "proxy currencies" (see Borgerhoff, Mulder and Sellen 1994).[15] Outside of human behavioural ecology, microeconomic models were exemplified by the economic defendability model, first applied in anthropology by Dyson-Hudson and Smith (1978), and further developed in the study of land property rights by Casimir (1992), which was partly based on the biological optimality matrix. According to this model, different forms of landed property are determined by the relative costs of securing exclusive rights on a territory (i.e., defense) and the benefits of having such exclusive rights. These benefits, on the other hand, depend on the predictability and spread of resources over the territory. A particular form of property, therefore, represents the optimisation of this cost/benefit ratio.

The second approach focused on the actual process of decision making with a view to building up its cognitive models (Orlove 1980: 246–247). Some studies using this approach were also related to the human environmental ecology school (e.g., Erdenebaatar 1996; Schareika 2001) and endeavoured to explain the decision-making process by formulating a set of decision rules, supposedly transmitted from one generation to the next through the process of enculturation. These studies, therefore, could be associated with the cultural models approach as described in the previous chapter. The theoretical assumptions upon which these studies were built have recently been criticised (see Ingold 1990, 1996, 2000, 2003). The majority of studies, however, were related to more traditional branches of anthropology and did not make explicit reference to cognitive cultural models of any sort (e.g., Beach 1981, 1990; McCabe 1994).

Neither approach has thus far resulted in a coherent explanatory model for nomadic pastoralists' decision making. This is probably because both approaches have significant shortcomings. Although the microeconomic models – based on optimality theory – can generate testable hypotheses regarding which factors influence nomadic pastoralists' behaviour, their abstraction from the decision-making process seriously hinders their ability to explain exactly how these factors work together to produce specific decisions. As theorists of the optimality models themselves acknowledge (see e.g., Smith 1983), these models do not explain how decisions are actually made, let alone predict them in detail. They would be better viewed as either shortcuts or credible guides to developing such an explanation. On the other hand, once an attempt to go beyond these abstract models and focus on the actual processes of decision making is made – as is done in the studies that adopt the second of the approaches described above – the diversity of factors that nomads would have to take into account and

15 Borgerhoff Mulder and Sellen define this term as "any outcome, such as maximisation, minimisation or variance reduction in a commodity such as food, herd size, labor or productivity. Currencies are only proxy measures of fitness, but must be proven to have (or likely to have) some effect on ultimate evolutionary fitness" (Borgerhoff, Mulder and Sellen 1994: 206).

the number of proxy currencies they would need to maximise with their decisions immediately become so unwieldy that developing a workable decision-making model – for example in the form of a decision-rules system – that would accurately reflect the cognitive operations supposedly performed by the nomads becomes an almost impossible task. However, as Ingold (1996, 2000) and others (e.g., Bourdieu 1977; Lave 1988) have pointed out, there is a significant problem here: indeed, this suggests that the people under research (in this case nomadic pastoralists) are capable of performing, in their everyday lives, much more complicated calculations with a much larger number of factors than we, in our university milieu, are able to model, let alone perform ourselves. On the other hand, as was convincingly demonstrated by Herbert Simon (1957), in the course of performing their everyday activities people usually fail to follow even relatively simple decision-making matrices, do not engage (at least consciously) in complicated calculations and use a rather limited number of very general rules of thumb (if any at all). Therefore, unless we fall back on radical psychological relativism and accept that the mental processes of nomadic pastoralists are fundamentally different from ours (which is doubtful – see previous chapter), they simply cannot have a decision-rules system that we would be unable to grasp.

Using our own data collected among reindeer-herding pastoralists, in the rest of this chapter we shall try to develop one possible solution to this conundrum. Before proceeding to this, however, another implicit feature that is common to both current and past attempts to create an explanatory model of nomadic-pastoral systems should be highlighted. Although it is quite clear that nomadic pastoralism involves both people and semi-domesticated animals, our examination of the existing literature indicates that only the former have usually featured as active agents in anthropological analyses of nomadic-pastoral systems. In the best of cases, nomads and their animals have been lumped together to produce a single unit of analysis (e.g., "coincident populations" as proposed by Neville Dyson-Hudson 1972: 19). However, even in these cases, the human population alone was assumed to be capable of determining the behaviour and development of the coincident populations, although it was believed to take into account the needs of the animal counterparts in the process. In all the models reviewed above, animals are assumed to have needs but no behaviour of their own to satisfy them. They are essentially seen as being dependent on their human masters, who are expected to perceive their needs, weigh them against a variety of ecological and non-ecological factors and, using their knowledge and will, provide for their "satisfaction" by moving the animals from one place to another, grouping and regrouping them, and so on and so forth. The general approach to modeling the functioning of nomadic-pastoral systems is roughly encapsulated in the following diagram (figure 2.1).

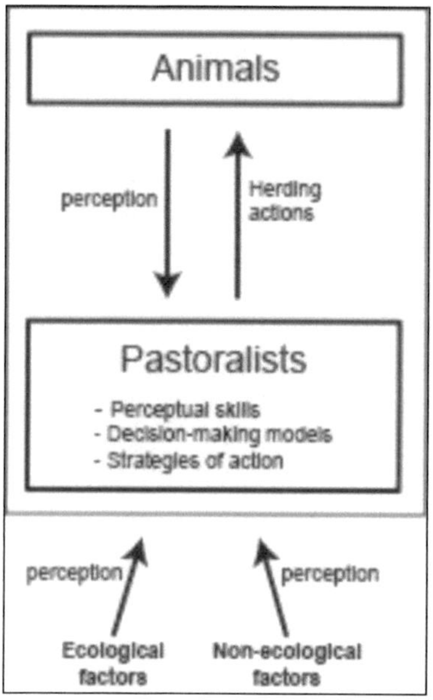

Fig. 2.1 A general model of the functioning of nomadic-pastoralist systems.

This general approach would appear to be based on the assumption that nomads can impose any decisions they wish on their animals. However, if some studies are read carefully, there are a number of observations that contradict this view. For example, Erdenebaatar (1996) mentioned that Mongol nomads were forced to make very frequent migrations between seasonal pastures because the animals tried to return to abandoned pastures, and that a high speed of movement was the only way to deter them. Similarly, Turner (1999) observed that cows became quite "picky" over grass species during some periods and were, therefore, calmer in more diverse pastures. Furthermore, Coppolillo (2000: 555) noted that large herds of cows tended to walk farther than smaller ones ("…ten steps without a bite"), and thus large herds moved faster. He also presumed that this could explain why large herds had longer migration routes than small herds, and also why nomads tended to divide their herds. However, these isolated findings seem to remain exceptions to the rule and unraveling the relationship between nomads and their herds remains largely vacant in the theoretical models.

In the following sections of this chapter, we will present fieldwork data from the two groups under study on the organisation of reindeer herders' activities. On the basis of this description, we argue that by paying more attention to the relationship between nomads and their herds, a new way of building explanatory models for nomadic pastoralists' movements can be achieved without assuming any superhuman calculating capabilities of nomadic pastoralists. This we hope will enable a more realistic account of cognitive tasks that reindeer herders face in the course of their day-to-day herding activities.

2.2 Maneuvering with reindeer herds: The practice of reindeer herding among the Izhma Komi

As was mentioned in the previous chapter, the pattern of nomadic migrations among the Komi reindeer herders is linear and vertical (from south to north). At present, the length of migration routes (*vörga*) can reach 650 km, which can add up to a total distance of up to 1300 km for the entire annual migration (from south to north and back). However, routes greater than 900 are known to have been used twenty-five to thirty years ago. The term *vörga* actually refers to a track left on the ground by the runners of reindeer sledges that follow the same path year after year. These tracks connect the winter pasturelands (*tövianin*) in the forest to the midsummer pasturelands (*gozhianin*) on the Arctic Sea coast, and they are easily recognisable in both the tundra and the forest even in the snowless period. Typically, Komi reindeer herders spend between four and four and a half months (November–March) each year in their winter pastures, about three to four weeks (from the second half of July to the beginning of August) in their midsummer pastures, and the remaining six and a half to seven months – that is, most of the year – travelling between the two along the *vörga*. It would, therefore, be logical to start our description of Komi reindeer herding practice by focusing on this long transition from winter to summer pastures.

The usual layout of a nomadic camp. The conical nomadic tent (chom in Komi, mia in Nenets) surrounded by sledges is in the center. The net in front of the tent is a part of movable corral where transport animals are driven for harnessing. Photo: Mark Dwyer.

As one might guess from the description above, the *vörga*s of Komi reindeer herders run parallel to each other from south to north. A distance of 10 to 20 km. separates one *vörga* from another. A number of campsites (*chom mesta*) used by the herders during the migrations are situated along the track, usually on the tops of hills and in close proximity to sources of clean water (streams or lakes) and firewood ("islands" of forest or willow stands). Campsites are located at approximately 8-km intervals in the northern section of the tracks, and at 15-km intervals in the southern section of the tracks. Each individual migratory movement is usually conducted between two adjacent campsites. Normally, each *vörga* is used by two herding units (called "brigades") of reindeer herders: one follows the other along the track, trying to stay two to three campsites behind. The basic land-use principle is that two brigades that share a *vörga* must not pasture their reindeer on the same side of it; in other words, one brigade should pasture its herd to the left of the *vörga*, and the other to the right (see Figure 2.2.). Therefore, the territory each brigade can use for pasturing its animals is the 10- to 20-km-wide strip of land situated between its own migration track and the adjacent track to the east or to the west of it. Furthermore, it normally has to share this territory with one of the groups that migrate along this neighbouring track.

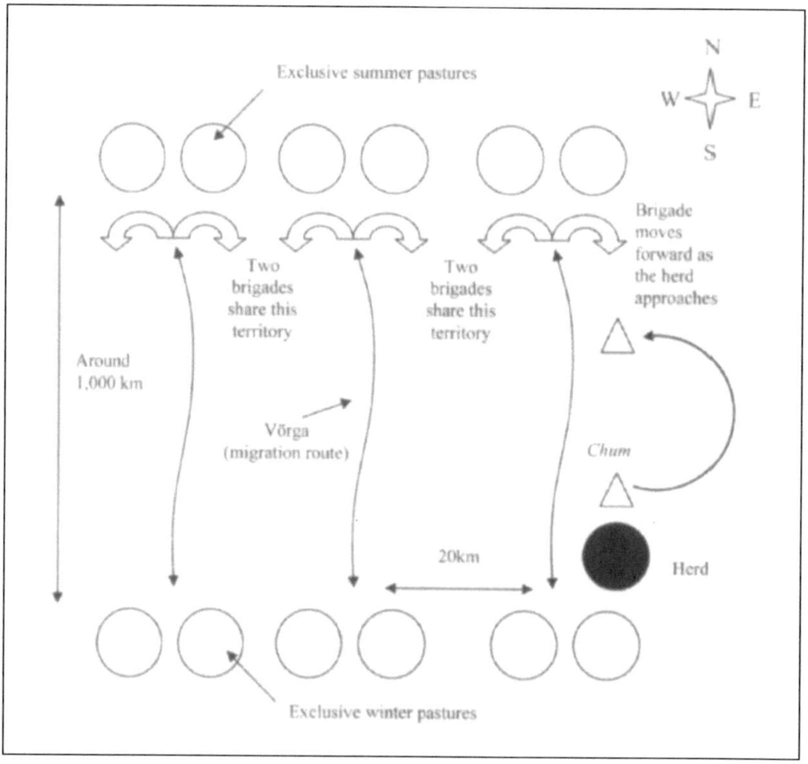

Fig. 2.2 Komi migration pattern and basic arrangement of pastures (from Dwyer and Istomin 2006).

Most herding groups divide their animals into two numerically unequal herds that are pastured separately. One herd, referred to as the *byk*, consists of castrated male reindeer used for transportation, while the other herd (*kör*) incorporates female and non-castrated male reindeer, as well as calves. The *byk* herd, usually numbering between two and four hundred animals, is always pastured in close proximity to the camp so that the nomads always have a means of transport at their disposal. The *kör*, which accounts for 80-90 percent of the total number of animals, can be 10 to 15 km from the camp. In spring and summer, when nights are short or non-existent in the Arctic, each herd is constantly watched over and controlled by a single reindeer herder (working a twenty-four-hour shift) with a reindeer sledge and between two and four herding dogs. In autumn, when dark nights reappear and make round-the-clock observation of the herd impossible, constant monitoring of the animals becomes restricted to the daylight hours. Obviously, pasturing the larger herd of productive animals (*kör*) is believed to be a much more demanding task than taking care of the small herd of transport reindeer (*byk*). For this reason the *kör* is usually controlled by experienced adult reindeer herders, while the *byk* is controlled by either less experienced reindeer herders (*iando*) or elderly herders whose age prevents them for doing the more demanding work. The work with the *byk* is, therefore, commonly considered to be a part of the socialication and education of young reindeer herders that allows them to acquire the necessary experience to deal with the *kör*.

A duty herder has three basic tasks: a) to keep the herd together and prevent it from dispersing or separating into smaller groups of reindeer (*palak*), some of which can easily slip out from under the herder's control; b) to protect the herd from environmental hazards such as predators and dangerous terrain;[16] c) to keep the herd within the boundaries of the territory his group can use for pasturing its animals and to prevent his herd from getting mixed up with herds of other brigades. In performing his activities related to these tasks, he has to bear in mind one general principle that plays an important role in the organisation of herding processes among Komi on the transitory land between winter and midsummer pasturelands: the larger herd (*kör*) should always stay "behind" the camp. This means that the herd should be to the south of the camp when the herders migrate northward (in spring and the first half of summer), and to the north of the camp when the general direction of migration is southward (in the second half of summer and in autumn). This principle is important for several technological reasons that will be explained later. Now it is enough to note that this principle plays a leading role in determining the schedule of the nomads' migrations. Indeed, in the process of grazing, the *kör* moves across the terrain away from the already grazed land toward the fresh pasturelands. Because the territory each brigade can use for pasturing is limited by its *vörga* and the *vörga*s of the neighbouring brigades to the east and west, this movement toward fresh pastures means movement northward in spring and southward in autumn. In other words, if the "herd-behind-the-camp" principle is obeyed, the herd always moves toward the camp. Once it gets close to the camp, the herders undertake a migration to the next

16 Dangerous terrain includes steep slopes, deep streams and bogs where reindeer can fall or drown. It also includes a thermokarst called *tabey* in Komi, which occurs as a result of the wet soil substance turning into a gel under a thin upper layer due to a mechanical impact. When reindeer cross *tabey* they can easily sink into the soil and die.

campsite, and the distance between the camp and the *kör* herd is re-established (see Figure 2.2.). The movement of the *kör*, then, automatically triggers the movement of the herders. On the other hand, too frequent migrations are not always in the interests of the herders, as they exhaust the draft reindeer and leave too little time for important subsistence activities, such as processing hides, sewing clothes, building and repairing sledges, fishing, and so forth. For these reasons, one further important task of the duty herder is to regulate the speed of the herd's movement in order to prevent it from reaching the camp too soon. This decreases the overall frequency of the camp's migrations.

Effective performance of these tasks depends mainly on the duty herder's ability to perceive and react to the herd's behaviour. Hence some initial account of this behaviour is needed here (a more detailed account will be given in chapters 6 and 7). The aspects of this behaviour that are the most relevant to the realisation of the tasks faced by the duty herder are the pace and the direction of the herds' movement, and the tendency of the animals to disperse over the terrain. These aspects are influenced by two chief factors: the communal instinct of the animals and the competition among animals within the herd (Baskin 2009; Skogland 1989, 1991). The former factor is what actually makes the whole enterprise of nomadic reindeer herding possible. An average reindeer has a natural urge to be in a herd, surrounded by a certain number of conspecifics. When there are no or very few other reindeer around, the animal becomes anxious and begins to move fast and in a rather complicated manner across the landscape in order to find and join a larger group of reindeer. This instinct ensures that a number of animals will normally stay together as a herd and can be dealt with as a single unit by a reindeer herder. The communal instinct of individual reindeer continues to work within the herd, causing a constant movement of animals from the borders of the herd toward its center (Baskin 2009). This tendency of reindeer to concentrate in big groups, however, immediately creates the second factor affecting the behaviour of the reindeer herd as a single unit: the increased competition for forage between individuals within the herd. This competition forces the individuals in the center of the herd to move rapidly in search of forage as they constantly try to overcome one another (Baskin 2009; Skogland 1989). To a certain degree this factor offsets the effects of the individual animals' communal instinct – the movement from the periphery of the herd toward its center – and can increase the dispersal of the herd. More importantly, however, the simultaneous impact of both factors is the primary mechanism inducing the constant movement of the herd over the terrain (Baskin 2009; Skogland 1989, 1991). As is clear from the description of this mechanism, both the speed of the movement and the rate of the dispersal of the herd directly depend on the abundance and quality of the forage in a given location. Indeed, the more abundant is the forage and the higher its quality, the less pronounced is the competition between individual reindeer in the herd. The movement of individual reindeer from the periphery to the center of the herd, then, becomes the most pronounced pattern of the animals' behaviour, and the herd tends to remain coherent and move slowly. Conversely, a decrease in the quantity or quality of the forage immediately causes an increase in competition, which increases the speed of herd's movement and ultimately leads to the dispersal of the herd. It should be added here that the dispersal of the herd and the speed of its movement tend to be inversely related to each other: the quality of the pasture being equal, a herd that is forced to stay more coherent (for example as a result of the herders' actions or other factors to be described below) tends to move faster than a herd that could freely realise its tendency to disperse. The reason for this effect is also easy to understand: the dispersal of the herd is a factor that decreases the competition between individual animals and, therefore, decreases the speed of their movement.

The amount and quality of forage, however, is just one factor affecting the dispersal of the herd and the speed of its movement. Other important factors include the accessibility of the forage (this is especially important for winter grazing, when the forage may be abundant but inaccessible to the reindeer due to an ice shield on the ground), the temperature, and the intensity of harassment by mosquitoes (in the summer months). Thus the high temperature and the harassment by mosquitoes in the summer decrease the herd's grazing activity and, consequently, the competition between individual reindeer for forage. Under these conditions, the animals tend to group together and form a coherent herd. On the other hand, the herd's movement tends to speed up significantly in these conditions because reindeer try to counterbalance the impact of heat and mosquitoes by running rapidly against the wind and/or toward the tops of hills or ranges. The direction of the wind is also an important factor in determining the direction of the herd movement when temperatures are low and mosquitoes are absent. However, in these conditions, the abundance and quality of forage becomes equally or even more important (Baskin 2009). In addition, the direction of the herd's movement under normal conditions can also be affected by landscape: reindeer normally tend to follow streams and natural depressions, as well as tracks left by other groups of reindeer. Finally, the coherence of the herd can be affected by factors that prevent the animals from having visual contact with each other. For example, if a group of animals gets separated from the rest of the herd by a hill, dense forest or a stand of bushes, it can start to behave as a separate herd and, furthermore, can easily get lost because such small groups of reindeer tend to move very quickly and cover great distances while looking for a larger herd to join. In order to finalise this short account of reindeer behaviour it is necessary to add that under normal conditions – that is, when the herd has sufficient forage and is not affected by high temperatures and/or mosquitoes – reindeer tend to follow a rather strict four-hour activity cycle: two hours of active movement and grazing are followed by two hours of relative passivity and ruminating (Baskin 2009; Skogland 1989, 1991). The presence of this cycle is a good indicator that the herd has enough resources in the given locality and can graze normally. On the other hand, factors that negatively affect the process of grazing tend to cause an immediate disruption of this cycle.

Given these clear patterns of reindeer behaviour, a herder performing his duty has two possible strategies to go about the realisation of his tasks. First, he can influence the behaviour of the herd directly by travelling around it on his sledge, frightening the animals (by shouting or ordering his dogs to attack them) that try to separate from the herd or go in an inappropriate direction, and forcing them to return to the herd. In this way the dispersed herd can be rounded up again; a group of reindeer that has become separated from the herd can be returned; a herd moving too fast or in the wrong direction (toward dangerous terrain or in the direction of another herd) due to mosquitoes can be stopped and turned back in its tracks. Secondly, the herder can attempt to change reindeer behaviour by playing with the factors that cause it. This normally involves moving with the herd to a new location. For example, if a herd tends to disperse in a particular place (e.g., as a result of intensive grazing in one location), one solution can be to move to another place that has more vegetation or vegetation of a different kind. If the herd moves too fast (in the case of intense heat or mosquitoes), it can be moved to a windier or cooler place. In addition, the factors influencing reindeer behaviour can be balanced against one another through movement. For example, taking the herd to a place rich in forage can decrease the speed of its movement (and make the tasks of a duty herder easier to perform) even on a warm, mosquito-ridden day. Moving the herd (or allowing it to move itself) to a new place is also the most efficient way to avoid dangerous terrain, as well as to decrease the risk of herd mixing. It should be remembered, however, that these maneuvers are rather restricted in space as the description of the land-use principles in the beginning of this section suggests. In fact, this maneuvering is always directed roughly toward the camp, as the land in the opposite direction (from where the herd had just come) would already have been grazed and moving back would increase the risk of herd dispersal.

It was observed that all Komi herders used both types of pasturing strategies. However, interfering with (*vörzödny*, lit. "touching") the herd too often to round it up, stop it, or change its direction of movement is inadvisable, as this disturbs the process of reindeer grazing and exhausts the duty herders' draft reindeer and dogs. The real art of reindeer herding primarily consists of accomplishing the herding tasks (i.e., keeping the herd coherent and within the delineated herding area, as well as protecting it) with minimal direct influence on the herd using a sledge and dogs, and minimal exploitation of the territory between the herd and the camp in the process of maneuvering.

In order to get some idea about how this outcome is achieved in practice, a description of concrete situations frequently encountered by a Komi reindeer herder on his duty in the *kör* herd can be helpful. After the duty herder comes to the herd in the morning and takes responsibility for it, he has an expanse of land in front of him that is limited by the two *vörga*s on the east and west, and by the present position of the herd and the present location of the camp to the south and north. On this territory his herd has to "live" the entire time until he is relieved by the new duty herder the next morning. In fact, if the herding group does not plan to move to a new campsite on the following day, the territory is even more restricted because the herder has to leave enough fresh pasture "behind" the camp for his colleague, who will deal with the herd the following day. Most reindeer herders start their work by mentally dividing this territory into two parts – one for grazing the herd during the daytime (*lun kezhlö*) and the other for keeping the herd during the coming night (*voj kezhlö*). Because, as was already explained, the herd is generally moving toward the campsite, the territory to be reserved for the night has to be closer in to the camp than the daytime territory. The task, then, becomes to draw a boundary between them at an appropriate place. In doing so, the reindeer herder exercises his knowledge of the probable behaviour of the herd – e.g., the speed of its movement and the rate of its dispersal – on particular parts of the territory, taking into account the expected weather conditions and their change during the day. The herder uses these expectations to plan his probable herding strategies and to get a sense of the probable work load. For example, the herder can decide to exploit a smaller territory or the territory with less or lower-quality forage for the day, and leave the larger territory or the territory with more forage for the night. This means that on the daytime territory the herd will move fast and will tend to disperse and, therefore, the herder will have to exert a lot of direct influence upon it in order to keep it within its predetermined limits, prevent the reindeer from escaping across the borders of the herding passage, and prevent their dispersal and separation into smaller groups. This will demand a lot of work from the herder, as well as from his draft reindeer and dogs. On the other hand, once the herd is settled on the larger and higher quality nighttime territory, the speed of its movement and its tendency to disperse will be greatly reduced, and the herder will be able to rest without having to exert much influence over the herd. The herder can further divide the daytime territory into smaller parts to be used during different periods of the day and, therefore, to further structure his anticipated herding strategies and the related workload.

Then the herder proceeds to the realisation of his herding plan. Normally, he starts by rounding up his herd and then giving it a direction of movement in such a way that the herd will start moving across the first of the delineated parts of the territory. Having done this, the herder takes a position on a nearby hill from which the grazing herd can be easily observed. If the delineated territory is large enough and the weather and mosquito conditions are not a concern, the herd will make its way across the territory, stopping once or several times for a two-hour long rumination before reaching the limits that were delineated by the herder. In the likely case that the herder thinks the herd should stay inside the territory longer, he will turn the herd back on its tracks, forcing it to go through the same territory again. As the forage on the territory has already been depleted, it will take much less time for the herd to accomplish its second pass through the territory, and its dispersion during this period will be considerably greater. It is quite probable that the herder will have to round the herd up several times in order to keep its dispersion within manageable limits, which will further increase the speed of the herd's movement. Theoretically, after the herd's second pass through the territory has been completed, the herder can try and turn the herd back on its tracks again and try to use the same territory for the third time. In this case, however, the amount of work needed to prevent the dispersal of the herd will increase further and, at some point, will become unmanageable for the herder. At this point, the herder will have to allow the herd to leave the delineated territory and start grazing on a fresh patch of land. According to the herders, the clearest sign that it is time to move on to a different pastureland is the disruption of the grazing–ruminating cycle of the herd: if the herd passes through a territory without a single stop for rumination, then keeping it on this territory is a bad idea – it will be too exhausting both for the herder with his draft animals and dogs and for the herd. However, the herders insisted that a good herder would never keep the herd on a territory until the disruption of the cycle. This is achieved through his skill in planning the usage of the land in such a way that the herd stays coherent and moves slowly by itself during the entire time of his duty, thereby requiring minimal direct influence. "In the hands of a good herder," the saying goes, "the herd more or less pastures itself." Such artful planning, of course, demands thoughtful observation and evaluation of reindeer behaviour, excellent ability to envisage its nuances in relation to the territory, and the balanced application of both direct influence on the herd and spatial maneuvering.

This ability to plan and maneuver is particularly important during the reindeer calving period in May. In this period, direct influence on the behaviour of the herd by rounding it up and redirecting its movements can disrupt the natural birthing process and, more importantly, the process of establishing bonds between the newborn calves and their mothers. This can result in the mothers' abandoning the calves, which leads to the death of the calves. To avoid this, the common practice is to keep invasive herding operations at a minimum during this period. The only way to achieve this is through the careful choice of appropriate calving grounds, i.e., those that are rich in forage and contain natural boundaries to prevent the herd from going off in the wrong direction, as well as a careful planning and use of these grounds.

Although this description of reindeer pasturing refers to the transition from winter to midsummer pasturelands and back, the reindeer pasturing in other periods is but a variation on this general scheme. In the case of the midsummer pasturelands, the most important difference is that the duty herder enjoys considerably more freedom in choosing territories for pasturing. The principle of always keeping the herd behind the camp does not apply, and the migration routes do not limit the territory available for maneuvering. The normal practice is to establish the camp in the southern part of the group's exclusive midsummer territory, where it remains for the duration of the midsummer period. The herd, on the other hand, moves in a big circle throughout the territory and, by the end of the midsummer period, returns to the proximity of the camp, triggering the start of the southward migration.

The practice of pasturing on winter pasturelands has more distinguishing features.

On the one hand, the speed of the herd's movement decreases greatly after the ground becomes covered with snow. At that time, the animals have to spend a lot of time and energy digging pits in the snow in order to reach the forage. For this reason, regulating the speed of the herd's movement is not an issue the herders have to deal with in winter. On the other hand, the amount of daylight is greatly decreased during this period (with only three to four hours of twilight from December through January), which makes the constant observation and direct influencing of the herd impossible. For these reasons the reindeer during this period are not so much pastured, but rather checked every day by one or two duty herders. The main factor the herders pay attention to is the dispersal of the herd, which, in contrast to the other periods of year, depends in winter not only on the abundance and quality of the forage, but also on its ease of accessibility to the animals, which is related to the snow conditions.[17] As it is difficult for the herders to react to the dispersal immediately, the risk of reindeer escaping from the herd is great. Just as in other periods of year, the herders' principal technique for dealing with the issue is to redirect the movement of the herd away from poor pasturelands. Again, this involves the ability to detect the dispersion tendency of the herd on an early stage and careful planning of the land use on the basis of expected reindeer behaviour, which, in this case, involves the evaluation of snow conditions, both present and possible in the future. Just as in the case of midsummer pasturelands, this planning is made easier by the considerable freedom the herders can exercise in choosing the best available pastures within the group's exclusive winter pasturelands.

2.3 Release and search: The practice of reindeer herding among the Taz Nenets

The practice of reindeer pasturing among the Taz Nenets differs from that of the Izhma Komi in two principal respects. First of all, in contrast to the long linear and longitudinal migration pattern of the Komi, the Nenets move in a circular manner. Normally, each herding group (camp) has a restricted territory of migration of approximately 60 km by 60 km, within which it pastures its reindeer herd year-round. No established migration routes are followed within this territory. Furthermore, Taz Nenets consider it unwise to use the same bits of this territory as seasonal pastures two years in a row,

17 These conditions include not only the depth of the snow cover (snow deeper than about one meter can render the fodder inaccessible to most reindeer), but more importantly, the quality of the snow. Frequent fluctuations in the temperature around the freezing point during the autumn can cause the snow surface to melt, which then freezes again and gets covered by the fresh snow. This can produce layers of ice under the snow (*tagös* in Komi) that are difficult for reindeer to penetrate. The worst condition is related to the formation of the ice cover directly on the ground. In this case the lichens – the main reindeer fodder in winter – get frozen into the ice cover and becomes completely inaccessible to the animals (*ii mu berdas* in Komi, lit. "ice on the ground").

let alone in two different seasons of the same year (as is the case among the Komi in their migration from winter to midsummer pasturelands and back). To avoid overexploitation of pastures, Taz Nenets reindeer herders generally change their migration routes from year to year.

Secondly, the Nenets pasture all their animals in a single herd, and this herd is not constantly monitored and controlled by a duty herder. Rather, the animals are left to graze unobserved most of the time. The herders round them up once a day in winter and twice a day in spring and autumn, and drive the entire herd to the campsite to check whether all animals are present, to select fresh animals for harnessing, and eventually to slaughter some animals for food and hides. After that the herd is driven away from the campsite and allowed to disperse over the pastureland again until the next roundup. Temporary deviations from this general pattern occur in the period of reindeer calving in May, as well as in the period when mosquitoes are abundant (end of June to the beginning of August). During the calving period, pregnant and newly calved female reindeer (*iahodei*), together with their newborn calves, are pastured separately from the rest of the herd (*ty*) and kept under the constant observation of a duty herder working a twenty-four-hour shift. The rest of the herd, however, is left to graze unobserved, and is just rounded up once a day. During the period of intense mosquito activity, the entire herd is observed and controlled constantly by two duty herders.

Even in these periods, the manner in which the monitoring is performed differs from that of the Komi. In the calving period, the herder normally does not interfere with the calving herd in any way: the danger of frightening the pregnant females and increasing their risk of miscarriage or calf abandonment is usually offered as an explanation for this. The duty herder simply observes the slow-moving *iahodei*, identifies and helps any cows that have problems giving birth (these animals usually stay behind the herd as it moves), collects the calves that die after birth or were abandoned by their mothers, and returns the females that try to disperse (usually those that have still not given birth to their calves) to the herd.

During the mosquito period, as was mentioned in the previous section, reindeer tend to stay together in a coherent herd that moves constantly and very quickly, most often against the wind. The Nenets explain that if the animals were allowed to move freely in this period, they could go so far away from the campsite that the herders might not be able to find them, round them up, and bring them back to the campsite in a timely manner. The main task of the two duty herders controlling the herd during this period, therefore, is to prevent the animals from wandering too far away from the campsite. This is achieved by controlling the direction of the herd's movement so that it rotates around the campsite as it is being attacked by the insects. If the mosquitoes temporarily disappear – which often happens when temperature drops significantly during the night or if the wind picks up – and the herd starts to disperse (signifying that the reindeer have started to look and compete for forage again), the herders immediately stop influencing the herd in any way and simply observe it until the insects reappear. No maneuvering with the herd to ensure its appropriate behaviour is evident during this time.

The herding activity that occupies Nenets herders during the great majority of the year, however, consists of finding and rounding up the herd that had been released from the campsite after the previous roundup, driving this herd to the campsite, and driving it away ("pushing it out") from the campsite again. This activity typically involves the joint effort of two herders in the winter and four in the spring and autumn, each with a reindeer sledge and team of two to four herding dogs. It also involves quite a lot of knowledge of typical reindeer behaviour to find the animals, as well as a special kind of planning, which, rather surprisingly, is not dissimilar to that used by the Komi reindeer herders described in the previous section. In order to understand how this activity is structured and realized, let us consider a real-life example that was recorded during fieldwork in October 2005.

On 27 October, shortly after the winter snow cover began to form on the ground but still before the freezing of the rivers, the camp of Nenets reindeer herders among whom we were conducting fieldwork migrated to a new location to the east of the river

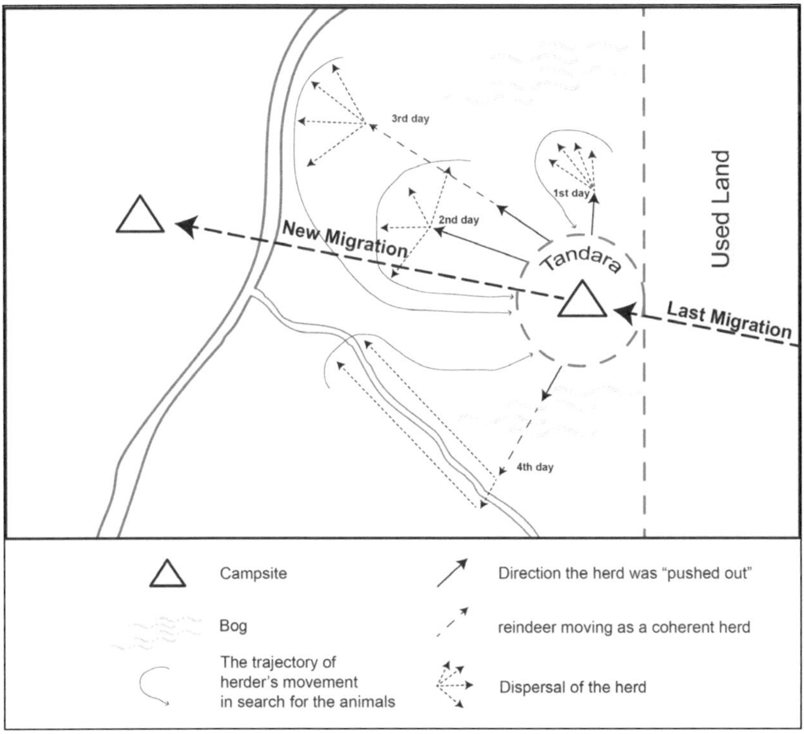

Fig. 2.3 Example of reindeer pasturing among Taz Nenets.

Limb'a-yaha (see fig. 2.3., eastern triangle). As is always the case with Nenets reindeer herding, this migration was executed in such a way as to establish the new camp outside the land already grazed by the herd when they were at the previous camp location (labeled "used land" in fig. ...). The herd was rounded up before the migration and driven behind the migrating caravans of sledges during it (also a standard practice of the Nenets that differs from the Komi). In the new location, a relatively large and flat area was chosen as the new campsite. The nomadic tents were erected in the center of this area, while the rest of it formed the *tandara* – the place where the reindeer herd, brought to camp after the daily roundup, could rest and find some forage before being released back to the tundra. Immediately south of the camp a frozen bog stretched about four kilometers up to the bank of a small but still unfrozen stream that flowed into the river Limb'a-yaha. The river itself was about ten kilometers west of the campsite. There was another frozen bog about five kilometers north of the campsite.

Immediately after the new camp was erected, the reindeer were "pushed out" from the campsite: three reindeer herders with leashed dogs frightened the reindeer, who were resting near the new camp, and forced them to run to the north, following them to the edge of the *tandara*. In this way the herd was released and given a northward direction of movement. According to the herders, the decision to push the herd out in a northerly direction was risky: it was based on the expectation that the herd would find enough forage in this direction and, therefore, would proceed rather slowly, not reaching the bog to the north by the time of the roundup the following morning. If the herd were to move faster than expected, it could have reached the bog, where there is very little reindeer forage in autumn, in which case it would have rapidly moved further to the north in search of better pasturage, increasingly dispersing along the way. This would have created tremendous problems for the herders trying to round the herd up the next day. On the other hand, the prevailing north-easterly wind that day and the relatively flat landscape to the north of the campsite virtually guaranteed that the herd, after being pushed to the north, would follow that direction as a coherent group and would not split into smaller parts each going its own way. This would make the task of finding and rounding the herd up the next day easier, assuming that the herd did not reach the bog. The herders weighed their options and took the risk, choosing to use the small territory to the north of the new camp first and leaving the larger territories to the west for the following days.

The following day the herders started to the north to round the animals up and found them in a coherent pack south of the bog – excellent proof that their expectations concerning the behaviour of the herd were justified. The herd was driven to the campsite and, after new draft animals for the herders' sledges were selected and one calf was killed for meat, pushed out again, this time in a westerly direction. By that time the wind had shifted to the north, and the herders expected that the animals would disperse widely by the time of the next day's roundup and would end up forming several groups, some moving to the west (the direction of the initial push), others to the northwest, and yet other groups heading north (against the wind). With this in mind, the next day the herders started to the northwest with the intention of making a big arc from the northwest to the west to round the dispersed groups up into a single herd. Contrary to their expectations, however, they found that the herd did not disperse much and was about halfway between the campsite and the river. This assured the herders that the pastureland contained a fair amount of forage. After rounding up the herd and returning to camp, the herd was again driven to the west that same day. This time it was expected that the animals would rapidly go through the land they had already grazed on the previous day and then disperse over the fresh pastureland beyond. The river would form a natural boundary that would prevent them from going too far away to the west. This expectation materialised, and the next day the animals were found on the eastern bank of the river.

Finally, on the third day the animals were driven southward. The expectation of the herders was that the reindeer would rapidly go through the bog, which was poor in forage, and reach the stream to the south of it. As stream valleys are usually rich in forage and reindeer generally like following natural depressions, the herd (or at least the majority of it) was expected to continue grazing along the stream. The herders explained that this expectation was particularly justified because there was no wind that day. Still, the herders acknowledged that a portion of the herd could cross the stream and proceed further southward, and they were worried that it would take some time to find them the next day. This could create certain problems because a migration to a new campsite was planned. On the next day, the herders departed in a westerly direction in order to find and round up the herd. Their plan was to intercept most of the herd in the lower part of the stream valley and then, if some of the animals were missing, proceed southeastward to find them there. Fortunately, all the animals were found in the stream valley. After they were brought to the campsite, the camp and the herd migrated to the western bank of Limb'a-yaha, to a new area that was outside the territory that had been used by the reindeer over the course of the previous days. The herders explained that staying longer at the campsite would not make much sense: all the suitable pasturelands near the campsite had already been visited by the animals and, therefore, the reindeer that had been released from the campsite could be expected to go quickly through them and to wander too far away from the camp by the time of the next roundup, making it time-consuming and demanding to find and drive them back to the camp. As the days were rapidly growing shorter, the risk of failing to find some of the dispersed groups of reindeer increased. It was indeed time to migrate further.

This example illustrates well the kind of knowledge, planning, and reasoning involved in the reindeer pasturing practices of the Taz Nenets on a day-to-day basis. A few words about the overall pattern of their migrations will serve to put this reasoning into a broader context. As the example above clearly demonstrates, each new campsite to which the Nenets herders migrate should have a large area of fresh pastures. Therefore, the longer the herders stay in one place, the further their subsequent move must be.[18] The migration routes of six Nenets herding camps over the course of several years and interviews with the herders suggest that there are two basic models of migration. The first model – which can be called the "double loop" – consists of two circles of migration: the winter circle (from November to the beginning of June), made in the forest and/or forest-tundra zone; and the summer circle (from June to November), made in the tundra. The circles are connected at one point, situated near the tree line, where the herders leave their winter equipment (e.g., reindeer hides to cover the winter tents and warm winter clothes) in June and pick it up in late autumn (November). The second model of migration, which we call the "three-circle" model, consists of three circles of movement: a winter circle from November to the beginning of June; a spring/summer circle from June to August; and a summer/autumn circle from August to the beginning of November (fig. 2.4.). These circles are also connected at the equipment storage place. In this model, the winter circle is also made in the forest/forest-tundra zone, the spring/summer circle follows watersheds covered in willow stands (*nero sä* and *paju-hoj*); and the summer/autumn circle is made in the open tundra (*lapta*). The herders return to the winter equipment storage place twice – once in the summer (usually in the middle of August) and once in late autumn – which allows them to minimise the number of items they carry during their movements. The concrete locations of the migration circles change every year: the same territory should not normally be used two years in a row because lichens (*n'ado* in Nenets) grow very slowly. This does not apply to the land along watersheds used in the middle of the summer. As the herders put it: "Reindeer do not need lichens in the summer because grass and leaves grow back every year."

The migration route is determined in the spring and is mainly related to where the equipment storage place is located. If situated between the watershed and the area of open tundra, on land that has not been in use for some time (a minimum of two years), the three-circle model of migration is chosen. The herders consider this model to be the most appropriate because it greatly reduces the burden on draft animals. It is, however, not always possible to perform because the open tundra areas immediately adjacent to the watershed may already have been used in the previous years. In this case, the double loop model is opted for, in which the herders go to the open tundra beyond the watershed and return in late autumn, performing only one summer circle.

The explanations herders gave for these migration patterns were sometimes contradictory and, most likely, simplified. What is important, however, is that these explanations more often than not referred to the expected reindeer behaviour in

18 During the calving period the herders in charge of the calves and calving females try to move as little as possible because migrations put a great strain on newborn calves. This means that by the end of the calving period, the herd of calves and females is usually 15–20 km away from the main camp, which would have moved several times to enable the herd of male reindeer to be successfully pastured.

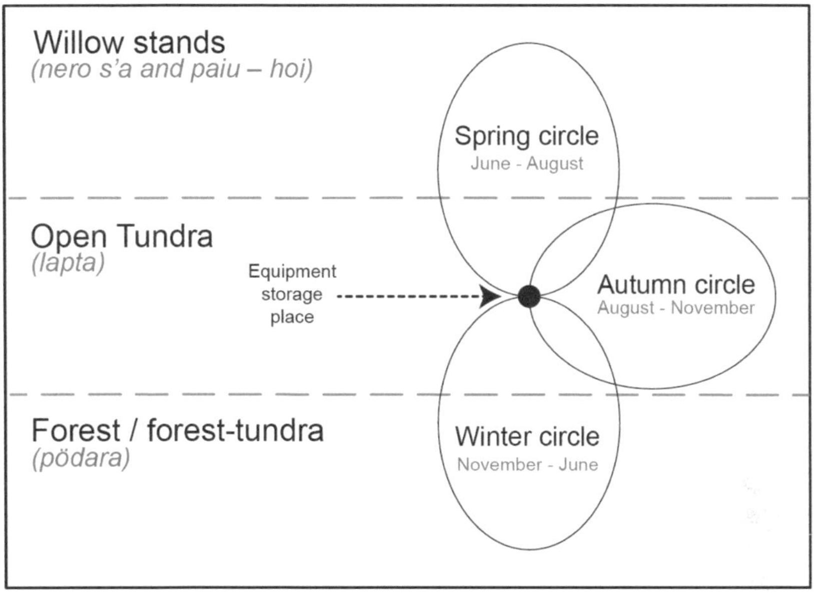

Fig. 2.4 Three-circle model of migration among Taz Nenets.

a particular territory and the workload needed to cope with it. Thus, the herders explained that making the winter circle in the forest/forest-tundra zone is mainly due to its softer snow cover, which makes it easier for reindeer to dig out the lichens and, in turn, means a more cohesive herd and limited herd dispersal. Another popular explanation was the greater availability of firewood in the forest/forest-tundra zone. They explained that willow stands do not make good winter pasturelands because it is difficult to look for dispersed groups of animals and round them up: indeed, the animals are difficult to see there from a distance, and traveling through dense bush stands on a reindeer sledge is not always possible. On the other hand, reindeer do not normally run fast through the willow stands, and the stands provide cooler pastures and thus some relief from mosquitoes. This makes the willow stands very attractive places to pasture the herd during the one month it is under the constant surveillance of the duty herders. Herders who attempted to pasture reindeer in the open tundra during the mosquito period often lost their herds because their draft animals became exhausted from the heat, mosquitoes, and constant workload, which limited the herders' ability to find and round up dispersed animals. Similarly, herders give two reasons for moving into the open tundra after the mosquito period: firstly, dispersed groups of reindeer are easier to find and round up in the open tundra than in the forest and willow stands; secondly, there are fewer mushrooms in the open tundra than in the forest or willow stands. Reindeer are very fond of mushrooms (which start to grow in the tundra in August), and disperse extensively in search of them.

Even this brief comparison between the Taz Nenets practice of reindeer pasturing and that of the Izhma Komi illuminates a number of rather important differences. However, it seems to us that there is also a striking similarity between them that is of particular importance as far as the main research question of our work is concerned. Indeed, it would appear that in both cases, the main factors influencing both the tactical and strategic behaviour of the herders are (a) the need to cope with and react to reindeer behaviour; and (b) to do this in such a way as to retain control over the animals while minimising the effort and resources required to do so (in terms of time, human labor, and draft reindeer and dog power). The main distinction between the two groups is that the Komi endeavour to control their animals constantly, while the Nenets, for most of the time, round up and release the reindeer regularly. Furthermore, in both cases the behaviour of the herders seems to be guided by the same cognitive ability to perceive, decipher and predict the behaviour of reindeer in relation to a wide set of biotic and abiotic factors, such as the quality and quantity of forage, the presence or absence of mosquitoes, temperature, snow cover, and so forth. It should be further observed, however, that despite knowing that reindeer behaviour is influenced by the quality and quantity of forage, the herders in both groups seem to make their herding decisions in relation to reindeer behaviour per se, rather than any assessment of forage. This observation, in our opinion, enables one to view the functioning of reindeer herding pastoralism in a new light, as we shall demonstrate in the next section.

2.4 Toward a new model of the functioning of nomadic-pastoralist systems

The ethnographic material presented in the two previous sections demonstrates that the functioning of reindeer pastoralism as a system cannot be understood in any detail if herders are considered the only active agents determining the behaviour and development of the coincident human and reindeer populations. This approach can, at best, lead to a partial understanding of a very general pattern of reindeer pastoralism. However, if the focus is placed on actual practice, it immediately becomes clear that it results from the relationship that exists between the herds and their herders. Consequently, in order to develop an explanatory model for reindeer herding, the behaviour and actions of the herders and their herds need to be separated, and the relationship between them focused upon.

The idea that the relationship between herders and their herds is an essential determinant of the parameters of a nomadic-pastoralist system is not particularly new. A number of accounts of this kind can be found in the literature on reindeer herding nomads. For example, in the second half of the nineteenth century, Maksimov (Maksimov 1984) argued that the Nenets did not actively pasture reindeer, but rather followed them. While this observation acknowledges the importance of reindeer behaviour, it still posits the rather extreme notion that there is a single locus of decision making, which, contrary to the more common practice of attributing all decision-making power to the herders, rests with the animals. This extreme position is strongly opposed by some scholars (see Beach 1990; Ingold 1986, 1988). For example, Beach argued that:

> Although the nomad's common explanation of his own movements is, "the reindeer decide," this is a truth with modification. There is great flexibility in the decisions that the reindeer can adopt as customary, and the reindeer nomad greatly assists in helping the reindeer make up their minds. [...] In effect, the reindeer come to decide what the herder (within bounds) has decided for them (Beach 1990: 269–270).

In order to achieve this, Beach argues that:

> It is vital that the reindeer herder know the possible range of behaviour of the reindeer under different climatic conditions. He must be able to predict reindeer behaviour during the shifting interplay of climate, grazing and landscape, and must adapt both his herding and his husbandry to these conditions. (Beach 1990: 268)

Paine (1994a: 31) put essentially the same idea in even stronger terms, arguing that "reindeer pastoralism rests upon the successful deciphering of herd behaviour by herders." However, up to now no attempts have been made to develop an explanatory model for reindeer herding nomadic pastoralism based on this idea.

We believe that such a model can be built by incorporating the relationship between herders and their herds through the concept of "herd control." This concept includes ensuring the herd's cohesion and manageability, protecting it from predators and dangerous terrain, and preventing the herd from becoming mixed with other herds. As demonstrated by the material presented above, herd control can be regarded as a product of the interplay between reindeer behaviour and the actions of the herders. During the process of this interplay, reindeer behaviour can be modified by the herders' actions, but by no means is it determined by them. Importantly, the material demonstrates that in their dealing with the animals, the herders, understandably, aim to act in such a way that expenditures (in terms of time, labor and draft reindeer and dog power) related to their actions are kept to a minimum. Hence, herd control can be best understood as a result of the herders' attuning their actions to reindeer behaviour so as to ensure the herd's cohesion, manageability and safety with minimal expenditure of resources. There appear to be various ways of achieving this. The Komi, for example, constantly monitor their reindeer, while the Taz Nenets leave the animals free to graze on their own most of the time, but conduct regular roundups.

Ecological factors influence the reindeer pastoralist system primarily through their impact on reindeer behaviour, which is very sensitive to even the slightest change in the environment. Thus, in their herding practice, the herders do not need to assess continuously an incalculable number of environmental factors and to act on the basis of this assessment, as most of the microeconomic and some decision-making models of nomadic pastoralism mentioned in the first section of this chapter implicitly or explicitly claim. Rather, they generally attune their actions to environmental variability by responding to changes in reindeer behaviour alone. For example, the amount and quality of forage is one of the basic factors influencing a reindeer herding system, particularly the migration pattern. However, from a reindeer herder's point of view, migrations are not determined by whether or not a pasture has become "exhausted"; rather, the timing of migrations and the placement of campsites are determined (albeit not solely) by the degree of effort that is required by the herders, their draft reindeer, and their dogs to keep their livestock under control on a given pasture. In short, the herders move when reindeer no longer want to stay on the pasture. The abundance of forage (a biotic factor) influences the "willingness" of reindeer to stay on the pasture or to move to another pasture: the less abundant the forage, the more willing the reindeer are to move. Herders are completely aware of this and take it into account in their strategic planning. However, in their actual behaviour, even if it represents to a certain degree the realisation of these plans, the herders react to the behaviour of reindeer rather than to the abundance or lack of forage.

Thus far we have discussed only how the reindeer herding system is influenced by the ecological requirements of the reindeer. However, the requirements of the reindeer herders themselves are obviously also important determinants of the system's functioning and reflect mainly the non-ecological factors influencing it. Some of these include the need of reindeer herders to exploit the resources that are available in the natural environment (e.g., water, fish, game, and wood for fuel); the herders' need to trade with the settled population as well as with other nomads, which is possible only in certain places and times; the need to meet up with the transport (e.g., helicopters) that is irregularly sent to them in order to bring supplies, pick up reindeer meat, provide medical care, give a lift to nomads going to the villages, and so forth; and the broader social needs of the herders (e.g., to seek companionship, visit relatives, relax in a warm flat and watch television), which are available mainly in the villages. The social and political settings and the various agencies involved in reindeer herding also have an important impact. For example, the reindeer herding enterprises can order the herders to bring their herds to special corrals at a given time in order to count the animals, select reindeer for slaughter, and provide veterinary treatment. The traditional system of land tenure binds Komi herders to established migration routes (*vörga*) and Nenets herders to defined (albeit changeable) tracts of land (*ya*), but the state can remove a patch of territory from the agricultural allocation (*vyvesti iz selskohoziaistvennogo oborota*), in other words, confiscate it from the herders and give it, for example, to an oil or gas drilling company (see Habeck 2005). Many other examples can be given.

The above factors influence the herders' interaction with their animals in ways that would often have been different if they had been made exclusively to maintain control over the herd with minimal expenditure of resources. These modifications in response to external factors, however, always increase the risk of losing control of the herd. This risk can be partially mitigated through an increase in workload (in terms of human, reindeer, and dog power). It seems to us that the herders, in their everyday practice, must constantly assess this risk and balance the costs of the additional workload needed to mitigate it against the costs of failing to respond properly to non-ecological factors of the kinds described above.

The following examples illustrate this point. On several occasions during our fieldwork, we observed that the Komi herders had to remain stationed at an agreed campsite to await the arrival of the school helicopter (either to drop the schoolchildren off in the tundra in June or to collect them and return them to the villages in September). However, it was routine for helicopters to be delayed for several days and even weeks. With such delays it soon became impossible for the herders to maintain control of the herd with minimal expenditure of resources through their customary techniques of maneuvering the herd because of the lack of fresh pastures behind the camp. However, considering that it was of overwhelming importance that the herding brigade await the helicopter, additional herders were sent to assist the duty herder to ensure that herd control was maintained by constantly rounding up the dispersing herd and turning it back on its tracks. The balance between keeping the herd in one place by rounding it up and turning it back on its tracks and the less direct maneuvering of it (which, as has been described in this chapter, is the distinct feature of Komi pasturing technology) became distorted. As keeping the herd under control behind the camp became ever more difficult, the herders exerted more and more pressure on the head of the brigade to allow it to move on.

On another occasion the Nenets were ordered by the herding enterprise to repair a damaged corral. However, to do this the brigade had to pasture its reindeer in an area deemed unsuitable. This caused the herd to disperse over a wide area, and the combined effort of as many as eight of the eleven herders present in the camp was required every day to locate all the reindeer, round them up and bring them to the camp. This made the repair work very slow. Eventually the repair work was abandoned, as avoiding a conflict with the enterprise over uncompleted work did not warrant the potential loss of reindeer.

All these considerations suggest an actor-based model of the functioning of reindeer herding systems that is rather different from the one presented in the first section of this chapter. The model presented in fig. 2.5. represents the functioning of a reindeer herding system as an interplay between reindeer behaviour and the actions of the reindeer herders aimed at achieving "herd control" with the minimum costs in terms of human and animal power. This interplay involves the herders' perception and deciphering of, as well as their reaction to, reindeer behaviour, and is influenced, on the one hand, by the impact that ecological factors have on reindeer behaviour and, on the other hand, by non-ecological factors that have an impact on the actions of the reindeer herders.

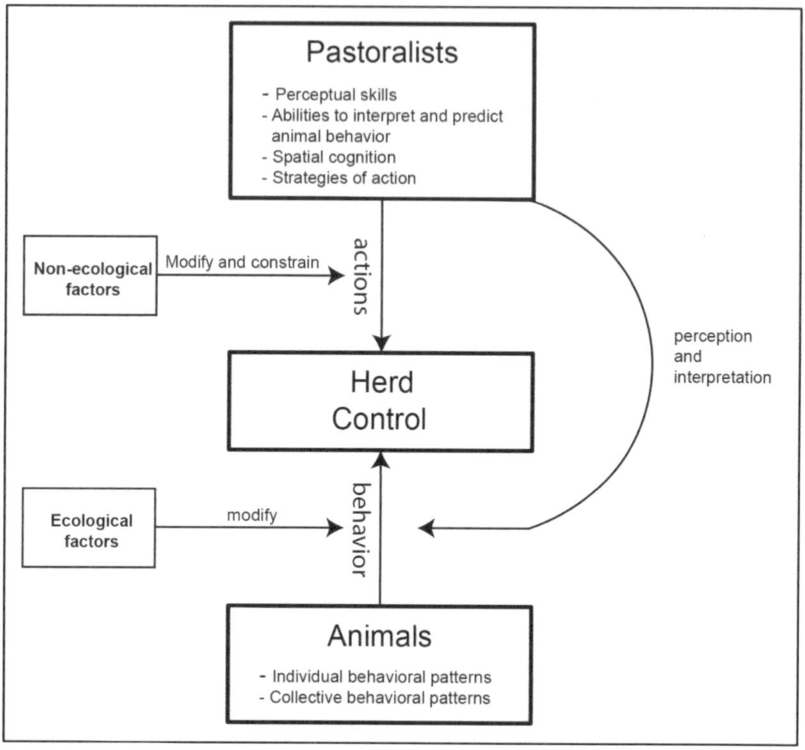

Fig. 2.5 Proposed actor-based model of reindeer herding.

This model has several advantages. First, the analysis of interrelations between the reindeer herders and their animals allows not only the determination of factors that influence reindeer herding, but also suggests how these factors work together. Secondly this model can work as a useful guide to building more specific and detailed testable explanations. Thus, it could serve as a guide when using optimality theory to analyse reindeer herding. When optimality models are applied separately to the reindeer herd (based, for example, on optimal foraging) and the herders' reaction to it (based on optimising manpower expenditure), the resulting behaviour can be regarded as the outcome of the interplay between these two optimisation strategies. If the model is correct, the hypotheses resulting from this approach should predict empirically observable behaviour better than those resulting from the model described in the first section of this chapter. Thirdly, and the most importantly for our purposes, this model can be used to clarify the cognitive tasks the herders routinely encounter in their herding activities.

Indeed, both the ethnographic material presented in this chapter and the model proposed suggest that in order perform to successfully as reindeer herders, Komi as well as Nenets have to solve two classes of cognitive tasks:

1. Perceiving, interpreting and predicting animal behaviour. The herders have to able to perceive the relevant details of animal behaviour, interpret them in terms of their causes and be able to envisage how this behaviour is likely to change with the change of relevant factors.
2. Planning and performing spatial actions. The herders should be able to plan and perform actions that will ensure herd control. As the ethnographic material suggests, these actions among Komi and Nenets reindeer herders always include a spatial dimension: they must maneuver animals (in the case of Komi), or "push them out" and then be able to predict their location (in the case of Nenets). Performing these actions demands a detailed representation of the environment, particularly the spatial distribution of features that are relevant to reindeer behaviour (landscape, type of vegetation, etc.). It also demands the ability to solve navigation problems successfully.

The mechanisms the herders employ to solve these two classes of tasks will be the main theme of the chapters to follow.

Finally, some general remarks concerning the proposed model and its application are in order here. It would, of course, be interesting to know how far, if at all, the principle of herd control can be applied to explain the herding decisions of other reindeer herding nomads, and of nomads in general. We are now entering a completely speculative arena as, in the absence of substantial comparative data it is difficult to make sound forecasts. It is probable that herd control is a major issue for all groups of reindeer herding nomads. However, it stands to reason that the degree to which herd control is an issue to the herders depends upon two factors: the level of reindeer domestication (which is largely determined by reindeer ecotype) and the size of a herd. One can imagine that herd control is less of an issue for small-scale reindeer herders (e.g., the Evenki, Tozhu Tyva, and some Khanty), and it may be that our model would not apply to them. On the other hand, herd control is probably an important issue for other large-scale reindeer herders (e.g., Chukchi, Saami, some groups of Sakha, etc.), for whom our model is likely to be applicable. It is even more difficult to judge with certainty how far (if at all) our model can be applied to nomads herding animals other than reindeer. Reindeer are known to be highly atypical among domesticated animals because the degree of domestication, which they probably acquired relatively recently (Baskin 2009; Igor Krupnik 1993; Pomishin 1990), is considerably less than other domesticated animals. As such, reindeer are considered semi-domesticated.This suggests that maintaining control over reindeer could be more problematic than over other animals. Nevertheless, as mentioned, there is evidence that southern nomadic pastoralists also have to deal with and react to specific animal behaviour. Studies have shown that some aspects of their migratory patterns depend upon this reaction, including the length of migrations (e.g., Coppolillo 2000; Erdenebaatar 1996; Turner 1999) and their direction (Turner 1999). However, we are not aware of any works that have systematically focused on this question. It is also evident that, in the case of southern nomadic-pastoralist systems, the herds more often than not contain several animal species with different behavioural patterns, and that there are different aspects of human–animal relations (e.g., milk yield) to which the herders would have to react.

Be that it as it may, the herd control model seems to be an adequate description at least of the reindeer-pastoralist systems of the two groups on which our study is focused. For this reason, it informs the discussion of particular aspects of reindeer herding cognition in the chapters to follow.

3 SPATIAL COGNITION AMONG REINDEER HERDING NOMADS: GENERAL MODELS AND CULTURAL PARTICULARITIES[19]

3.1 Spatial cognition and its functions in reindeer herding practice

The classic definition of the term "spatial cognition" is the "internal or cognitive representation of the structure, entities and relations of space; in other words, the internalised reflection and reconstruction of space in thought" (Hart and Moore 1973: 248). This definition is not very convincing because, as has been demonstrated by many studies (e.g., Golledge 1999; Kaplan 1973, 1976; Kitchin 1994; Lynch 1960; Moore and Golledge 1976), the internal representation of spatial relations between environmental objects is not easily distinguishable from the internal representation of other properties of the objects in the total environment (usually termed "environmental cognition" – Moore and Golledge 1976). The distinction between spatial cognition and environmental cognition in psychology and behavioural geography has emerged mostly because spatial cognition studies have focused mainly on mechanisms of place recognition and wayfinding. This is understandable, as wayfinding and place recognition are easily observable empirically and can be tested experimentally both among humans and non-human species. For this reason, a considerable body of literature exists on how spatial representations can provide answers to questions such as: Where am I? (e.g. Castner 1995); Where are the phenomena for which I am searching? How can I know where to go next from any given point in a layout? (e.g. Leiser and Zilbershatz 1989; Levine 1982); How do I know when I am lost? and How can I determine the distance between locations? (Montello 1991).

However, it has already been recognised by Lynch (1960) that wayfinding facilitation cannot be the only, or even the primary, function of spatial cognition. Spatial cognition is necessarily involved in any kind of spatial action, but the spatial action cannot be reduced simply to "targeted movement" unless we adopt the purely behaviouristic approach of animal psychologists. Any spatial action involves aim and purpose as well as complex relations to previous and following actions. In order to perform it, a subject needs to know not only where certain places or objects are, but also what they are, what properties they have, what the possibilities and/or dangers are in relation to his activities, and so on. In other words, spatial cognition can be viewed as only one of the aspects of environmental cognition, which is:

19 This chapter is partly based on a research article published in *Current Anthropology* (Istomin and Dwyer 2009).

> the awareness, impressions, information, images and beliefs that people have about environments… it implies not only that individuals have information … about the existence of these environments and of their constituent elements, but also that they have impressions about their character, function, dynamics and structural interrelatedness and that they imbue them with meaning, significance and mythical-symbolic properties. (Moore and Golledge 1976: xii).

From this point of view the primary purpose of spatial cognition is to *act as a spatial organizer of environmental information and experience* (Lynch 1960; see also Downs and Stea 1977; Golledge 1999). In other words, it provides a cognitive structure that spatially organises the information about and experience of environments. This structure exists along with other structures (for example, conceptual trees of different kinds), which organise the same information in other ways. This structured spatial organisation of environmental information enables a person "to predict the environment which is too large to be perceived at once and to establish a matrix of environmental experience into which a new experience can be integrated" (Stea and Blaut 1973, 227).

From the description in the previous chapter, one can see that such spatial organisation of environmental information is likely to play a very important role in supporting the reindeer herding practices of the two groups under study. Indeed, both maneuvering a reindeer herd (in the case of the Komi) and "pushing out" and locating the herd (in the case of the Nenets) essentially depend on the herders' knowledge of how reindeer are likely to behave in particular areas or locations around them. This presupposes knowledge of what these areas and locations are, how they are spatially distributed in relation to each other as well as to the current position of the herder and his campsite, and what the behaviour of reindeer is likely to be taking into account other relevant factors (season, weather, mosquitoes, snow conditions, etc.). This knowledge is indispensable for maintaining the herd control that is achieved primarily through influencing the spatial movement of the animals. Thus, on the basis of this knowledge, alone or in conjunction with their ability to perceive and decipher reindeer behaviour, the herders can solve the following tasks that seem to recur in their everyday practice:

1. imagine where the reindeer herd is likely to be at any given moment;
2. predict how the herd is likely to move and behave in the short and medium term;
3. decide on the best sequence of herding operations, such as the trajectory of maneuvering the herd during the duty period (Komi) or the direction of "pushing out" the herd from the campsite (Nenets);
4. find their way to the reindeer herd as well as to other locations they might need to visit (such as fishing spots, places to cut firewood, permanent settlements, etc.) and back to the campsite;
5. decide on the best place to move to and then to find their way there while performing movement (Nenets);
6. perform the planned maneuvering of the herd (Komi).

In performing these tasks, one of the most important cognitive tasks reindeer herders face is to create, maintain and update a detailed cognitive representation of a space expansive enough to allow effective pasturing of reindeer throughout the entire year. Furthermore, this representation should be structured in such a way that information about reindeer behaviour can be easily cross-referenced with it, allowing a further representation of a possible range of reindeer behaviour in different parts of the space represented. In this chapter, we begin to investigate how this task is realised and what cognitive processes are likely to be involved in it. Before proceeding to this investigation, however, it will be helpful to review the current models of human spatial cognition proposed in geography and cognitive psychology and the kinds of variation in them that culture is likely to bring about.

3.2 Human spatial cognition: Existing models and probable sources of cultural influence

The common term used in cognitive psychology and behavioural geography to refer to a cognitive structure or matrix that is capable of organising environmental information spatially is *cognitive map*. It is important to note, however, that this term, together with many others, is often used also in a more narrow sense to refer to one particular kind of architecture of such a structure (referred to as "mental map" in this book). In order to serve its function of organising environmental information spatially, the cognitive map (in the wider sense) should somehow encode (but not necessarily reflect) the spatial relations between environmental entities experienced. Indeed, these encoded spatial relations actually constitute this map as a cognitive structure. As far as cognitive maps of large-scale environments are concerned,[20] two possible modes of such encoding have been proposed and experimentally demonstrated (O'Keefe and Nadel 1978; Shemyakin 1962; Siegel and White 1975). The easiest and the most straightforward of these, usually termed "route (or route-based) knowledge" or "network knowledge" (Allen 1999; O'Keefe and Nadel 1978; Shemyakin 1962; Siegel and White 1975), is the mental representation of routes between entities. This can be achieved by a representation of a set of "temporal sequences of environmental features" (Allen 1999: 71), each of which uniquely specifies a certain route. The features

20 In modern cognitive psychology, three basic kinds of environment are distinguished on the basis of size and behavioural relevance (Montello 1991; see also Rizzolatti and Sinigaglia 2008): figural space, vista space (small-scale environment), and environmental space (large-scale environment). It is suggested that the spatial relations within each type are coded (represented) in different ways (Rizzolatti and Sinigaglia 2008). Figural space has the size of a particular object. Remembering the distribution of different things on one's desktop is a good example of its mental representation. Vista space (small-scale environment) can be seen in its entirety from one vantage point. Large-scale environment, which is the primary concern of this paper, cannot be seen in its entirety from any particular point within it.

referred to here, in the case of humans, can be landmarks (situated on the route or just visible from it), quality of surface (ascending or descending slopes), trails or paths, as well as combinations of all these in visual perspective. They can be represented (according to various theoretical models) in the form of a sequence of visual images (Allen 1999; Shemyakin 1962; see also Gell 1985) or so-called vistas (see Gibson 1986; Heft 1996 for definition and discussion; see also Ingold 2000). The temporal characteristic of the representation of the sequence should roughly correspond to the temporality of encountering the represented features in the course of actual travel. Route knowledge, then, encodes space as a set of routes (encoded by itself as sequences of representations of environmental features) between a set of objects (termed "places" or "nodes" – Golledge 1999).

From the viewpoint of psychologists and cognitivists, route knowledge has the great advantage of simplicity: both its learning and use (especially in wayfinding) can be modeled with reference to very simple psychological mechanisms. In fact, learning route knowledge should hardly involve anything more than visual memorisation, while navigation based on route knowledge can be easily modeled as a simple stimulus–response chain (Tolman 1948; Siegel and White 1975). However, spatial relations encoded by means of route knowledge do not preserve the actual Euclidean[21] dimensions of space. Thus, on the basis of route knowledge alone a person cannot determine, for example, the distance between two objects. He or she can only estimate the length of the shortest known route between them, which could be much longer than the actual Euclidean distance. In addition, route knowledge cannot indicate the actual position of the subject in Euclidian space, although it can specify his or her position on the route in terms of which part of the route has already been covered and which part is still left to cover. In other words, from the viewpoint of Euclidean geometry, route knowledge can answer neither "Where is a particular place?" nor "Where am I?", but only "How can I get from one place to another?" (Allen 1999; Castner 1995; Passini 1980; see also Ingold 2000).

However, as demonstrated as early as 1948 by Edward C. Tolman, an animal psychologist, route knowledge could not be the only form of spatial cognition among both animals and humans. In his renowned experiments (Tolman 1948), rats were initially placed in a maze with a food container situated at the end of a long curved path at some distance from the rats' home base. After the rats had become accustomed to travelling along this path to find the food, the maze was changed: the "usual" path to the food was blocked, while a number of other paths leading in different directions from the home base were opened. After exploring the "usual" path only to find it blocked, the majority of the rats chose a straight path that led directly to the container, despite the fact that the new direction was absolutely novel to

21 Here and further in the text, the term "Euclidean" refers to geometric attributes of and relations between points and figures in a two-dimensional plane as conceptualised in classic (Euclidean) geometry.

them (the "usual" path led them to the container in a roundabout way that changed direction several times). This led Tolman to conclude that "in the course of learning a maze, something like a field-map of the environment gets established in the rat's brain" (Tolman 1948:192). What Tolman meant was that the rats built up an abstract representation of the spatial relation between the home base and the food container in the form of direction and distance, rather than simply memorising the path between them. This representation made it possible for them to immediately choose a new route to the container when the old route was blocked. Tolman coined the term "cognitive map" in order to refer to this form of spatial representation, and this term, along with a number of others,[22] is still very much in use. Here, the term "mental map" is employed when referring to this form of spatial orientation, and the term "cognitive map" is reserved for the spatial representation involving several forms of architecture.

Tolman himself understood the cognitive (mental) map as an internal representation (model) of spatial relations between environmental objects in the form of directions (vectors) and distances in Euclidean space – much in the same way that the artificial map can be understood as a graphical representation of Euclidean relations between a set of environmental objects. This understanding is evident in most studies on spatial cognition performed after Tolman, regardless of the actual way in which Euclidean relations are represented in the mental map, and its possible similarity to the graphic representation exemplified by the artificial map is a matter of considerable debate (see Kitchin 1994 for a review). It seems reasonable, therefore, to accept the cautious definition of the concept by O'Keefe and Nadel, according to which the mental map is "a representation of a set of connected places which are systematically related to each other by a group of [geometric, i.e., Euclidean] spatial transformation rules, which permit generalisations and inferences beyond the specific spatial information gained through direct experience" (O'Keefe and Nadel 1978:86; see also Allen 1999:71).

Being agnostic about the exact way in which Euclidean relations may be represented in mental maps, this definition highlights instead the main cognitive property of this form of spatial cognition: its capacity to provide the basis for inference beyond direct experience. This fundamentally distinguishes the mental map from route knowledge. In fact, in contrast to route knowledge, the mental map readily answers the questions "Where am I (in relation to other places)?" and "Where is a certain place (in relation to where I am)?", while the information on how I can get from this to that place can be only inferred from these answers. The result of this inference can be a novel route never before experienced by the subject, as Tolman's rats demonstrated. This possibility is absent in the framework of route knowledge:

22 There are no fewer than sixteen terms used in the literature to refer to this concept (see Golledge 1999:xiv for a list, and Kitchin 1994 for an extended discussion). The terms "mental map" and "cognitive map" are, however, the most widespread.

the only way a subject who possesses only route knowledge could discover a new route between known places would be through a random search starting from one place (Allen 1999), i.e., by trial and error. This distinction is often used by animal psychologists as a decisive indicator of the presence of mental maps: "To be credited with a cognitive [in our terms, mental] map a subject therefore must be able to perform adequate new routes, that is, choose the most economical alternative path (such as a shortcut or a detour) under new conditions" (Etienne et al. 1999: 197–198).

It is commonly believed by psychologists, geographers, and cognitive scientists that human spatial cognition includes both route knowledge and mental maps, and that this fact is global (Reginald G. Golledge 1999).[23] The claim raised by anthropologist Tim Ingold (2000) that some, if not all, human indigenous communities have only route knowledge is unjustified (Istomin and Dwyer 2009). Indeed, as follows from the above discussion, if this were the case the members of such communities would be unable to shortcut.[24] More importantly, however, they most probably would be unable to organise their environmental knowledge and/or experience (whether in the form of internal representations or locally based "education of attention," as Ingold would probably put it) adequately and, therefore, their abilities for informed (or skillful) actions in the environment would be greatly hindered. For these reasons we agree with Golledge and colleagues (Reginald G. Golledge et al. 1983) that route knowledge as the only or even primary mode of spatial cognition can be found exclusively among persons with mild to moderate mental disabili-

23 It should be noted that apart from survey knowledge (mental maps) and route knowledge, the wayfinding process can also make use of so-called path integration (more commonly known as dead reckoning). Path integration is based on egocentric representations of Euclidean spatial relations between a subject and the locations visited. This can be achieved by keeping a mental list of turns he or she has made and distances he or she has traveled between turns. Alternatively, a traveler can have a representation of Euclidean relations between him or herself and the points in the environment and constantly update this representation in the course of his or her travel using some computation mechanism sensitive to the speed and direction of his or her movement (so called 'egocentric updating model', see Wang 2012). Reliance on path integration is currently believed to be the main method of wayfinding among insects and some rodents (Etienne et al. 1999; Gallistel 1993). Interestingly, all the cognitive abilities traditionally associated with mental maps (shortcuts, detours, ability to point to a location) can also be attributed to path integration (Collett and Graham 2004; Etienne et al. 1999; Gallistel 1993; Wang 2012). It has, therefore, been proposed by Wang (2012, see also Wang and Spelke 2002) that human wayfinding, like that of insects and rodents, can also be based on path integration rather than mental maps. Unfortunately, their evidence is derived from studies of humans pointing and moving in small-scale (room-size) environments. The application of their model to wayfinding in large-scale environments can be problematic: it is believed that the human capacity for path integration in large-scale environments is modest (Loomis et al. 1993, 1999). Still, path integration can play a role in human wayfinding as a part of a more complicated mechanism involving also other forms of enduring spatial representations. We will return to this point in the next chapter.

24 To the best of our knowledge, the existence of a human community consisting of members incapable of finding shortcuts has never been demonstrated.

ties while "locationally matched low socioeconomic mentally able persons quickly develop two-dimensional layout knowledge of both routes and the surrounding environment" (Golledge 1999: 10).

The relative roles of mental maps and route knowledge in human spatial cognition, as well as the relationship between them, are still matters of debate. Until recently, it was believed that route knowledge precedes survey knowledge both in ontogenesis (during the course of human development) and in microgenesis (during the course of learning a particular environment) (Hart and Moore 1973; Kulhavy and Stock 1996; MacEachren 1992; Moore 1976; Shemyakin 1962; Siegel and White 1975). The belief that route knowledge precedes survey knowledge in ontogenesis is based mainly on the Piagetian assumption that the capacity for abstract operations on mental representations is formed relatively late in a human's life (Piaget and Inhelder 1967). This assumption has been challenged by empirical studies that suggest that babies and young children can possess both route and survey knowledge of the environment (Acredolo 1981; Newcombe and Huttenlocher 2003; Newcombe 1997). The belief that people who start to learn a novel environment acquire route knowledge first and that survey knowledge develops much later as the experience of their environment increases (Delvin 1976; Garling et al. 1981; Hart and Moore 1973; Siegel and White 1975) is based on the assumption that route knowledge is a necessary prerequisite for the formation of survey knowledge (Shemyakin 1962; Siegel and White 1975). At least two theoretical models explaining the formation of survey knowledge based on route knowledge have been proposed (Golledge and Spector 1978; Siegel and White 1975). Both models assert that the cognition of a new space (territory) starts with learning routes (building route knowledge) between a set of objects or places. As the number of known places and routes increases, people start to retrieve the relative Euclidean position between them and, therefore, form survey knowledge (mental maps). These models have, however, also been recently challenged by experimental studies (Brunyé and Taylor 2008; Foo et al. 2005; Ishikawa and Montello 2006) that demonstrate that route knowledge does not necessarily precede survey knowledge, nor does it necessarily lead to the formation of survey knowledge. Rather, both forms of knowledge can be acquired independently of one another at the very beginning of the learning process.

Regardless of the order of acquisition of the two forms of knowledge, the formation of a mental map does not cause route knowledge to disappear. The subject mentally stores the representations of routes between places both inside and outside the regions covered by his or her mental map. These representations are frequently invoked to solve various cognitive tasks, especially those of wayfinding. As has already been mentioned, wayfinding based on route knowledge (i.e., route following) is much simpler from the cognitive point of view than inference based on a mental map. Therefore, subjects probably rely upon it whenever possible, whereas inference from a mental map is used only when route knowledge fails or provides an economically unviable decision (for example, if the known route to a place is too long or becomes unavailable).[25] Furthermore, an inferred route can certainly be visually learned and, therefore, a mental map can facilitate the acquisition of new route knowledge. On the other hand, route knowledge is cheap to use but expensive to store, and some researchers argue that one of the functions of mental mapping consists of eliminating the necessity to store a large and increasing number of route representations (Allen 1999; Garling et al. 1981). After a mental map is formed, the subject can act effectively in space even though only a limited number of the most frequently used routes are known. If the need for them arises, the new routes can easily be inferred from the mental map and learned in the form of route knowledge if they are to be used frequently. On the other hand, routes that are no longer used can be blotted out without many negative consequences (Garling et al. 1981).

The capacity to acquire each form of knowledge, as well as the speed of this acquisition, seems to depend on a number of idiosyncratic and social factors that differ between individuals (Ishikawa and Montello 2006). It was demonstrated, for example, that the formation of survey knowledge is likely to depend on sex. Women required a significantly longer time to form a mental map, and they less frequently used it for spatial orientation (Montello et al. 1999; cf. Lawton and Kallai 2002; Prestopnik and Roskos-Ewoldsen 2000). Indeed, according to Silverman and Eals (1992), the difference between the mental mapping abilities of men and women is related to the division of labor between them during the Pleistocene period, with men being involved in hunting and women in gathering. More importantly, the relative reliance on each of the two forms of knowledge seems to depend on a person's exposure to the environment and the extent to which he or she engages with it; general traveling experience would seem to further facilitate the acquisition and use of mental map (Maguire et al. 2006; Spencer and Weetman 1981; Spiers and Maguire 2006, 2008). As differences between persons in these idiosyncratic factors are likely to depend on their culture and way of life, the relative role of the two forms of knowledge in spatial cognition is likely to be one of the aspects where culture can "play variations" on cognitive universals.

Another source of cultural variation in the overall structure of spatial cognition is likely to be related to language (Levinson 1996, 2003; Wassmann and Dasen 1998; Widlok 1997). It has long been known that spatial relations between objects can be expressed linguistically either in relative or absolute frames of reference (see Levinson 2003: 25–34 for a review). In the first case, exemplified in English, for example, by the expression, "The glass is to the left of the bottle," the expression depends on the position and orientation of the speaker in relation to the objects. In the second case, exemplified by the expression, "The glass is to the north of the bottle," this dependency is absent. As Levinson (2003) explains, there can be also a third frame of reference

25 Recall the behaviour of mice in the experiments of Tolman (1948). The mice took the new route to the food container only after they tried the old one and found it blocked.

– intrinsic – exemplified by the English expression, "The tree is in front of the house." The expression given in this frame of reference does not depend on the position of the speaker unless, as is often the case, the position of the object is described in relation to him or her ("The tree is in front of me"). However, these expressions depend on the orientation of one of the objects in space and, most importantly, on the shared convention of speech of the speaker and the addressee on how sides (front, back, left and right) should be assigned to this object. Although the majority of human languages seem to allow expressions in all three (or at least more than one) frames of reference, there are languages for which this is not the case (Levinson 2003). More importantly, even if it were the case, not all linguistic frames of reference are equal in the frequency of their use in everyday speech. For example, among all European peoples the relative and intrinsic frames clearly prevail in everyday conversation, while the absolute frame is reserved for a limited number of situations. The great contribution that recent ethnolinguistic research has made consists of demonstrating that this way of expressing spatial relations is not universal. There are a number of groups whose representatives express spatial relations predominantly in the absolute frame of reference. The representatives of these groups, therefore, routinely use expressions like, "Take a cup from the southernmost box," "You have a fly on the eastern side of your face," or "Put your signature in the north western corner of this document" (Levinson 2003).[26]

It appears that relying on absolute linguistic frames of reference in everyday speech logically presupposes the possession of and reliance on extended spatial representations in the form of mental maps. Indeed, the equivalent of English phrases such as, "The school is to the left of the supermarket," or "Turn right from here to get to the town hall," would be coded in this frame as something like, "The school is to the north of the supermarket," or "Go south from here to get to the town hall." As Levinson pointed out,

To achieve such descriptions, one will need to know, or be able to recover, the angles subtended between any two places. In that sense a mental map of one's world with accurate absolute angles must be accessible. One place in that map that will be especially important will be the location of speaking [necessary to put together such phrases as, "The city hall is south of here" – *K.I., M.D.*]. As the speaker moves, his angular position vis-à-vis other places will change. (Levinson 2003: 125)

In other words, in order to rely on the absolute linguistic frame of reference, one needs to know at any moment how one is situated in relation to any object to which

26 It should be made clear that an absolute frame of reference would not necessarily be based on geographic bearings (east, west, north, and south), although this type of absolute frame is probably the most widespread. For example, the absolute frame of reference used in Bali is based on directions called *kangin*, *kauh*, *kaja*, and *kelod*. While *kangin* and *kauh* more or less correspond to east and west, *kaja* signifies the direction toward the mountains (away from the sea), while *kelod* is the direction toward the sea. The geographic direction of the *kaja* – *kelod* axis is, therefore, different in different parts of the island (Wassmann and Dasen 1998; see Levinson 2003 for further discussion).

one wishes to refer, as well as how these objects are situated in relation to each other in Euclidean space. A number of studies, mostly based on so-called pointing experiments (see section 3.3. below), have been conducted with people who rely on the absolute linguistic frame of reference. These studies demonstrated that these people indeed possess surprisingly large and detailed mental maps in comparison to people from modern industrialised nations and do constantly update their position in relation to the landmarks encoded in them (e.g. Haviland 2000; Levinson 2003; Lewis 1976). These results, among others, have been used to support the claim that the spatial cognition of people relying predominantly on an absolute frame of reference is different from that of people relying predominantly on relative and intrinsic linguistic frames of reference. Levinson (2003) argues, therefore, that "absolute" people are likely to have "oriented survey maps" (i.e., mental maps that encode the relative positions of objects in Euclidean space using a general geocentric system of coordinate axes, for example geographic bearings). "Relative and intrinsic" people, on the other hand, rely predominantly on route knowledge or non-oriented mental maps (i.e., mental maps that do not employ any general system of coordinate axes) (Levinson 2003: 273–275). Although there is still little empirical support for this argument, it nevertheless indicates that the structure of mental maps can also be a potential area of cultural variation in human spatial cognition.

Other studies inquiring into structure of mental maps have resulted in an important discovery of its hierarchical character (e.g., McNamara et al. 1989; Stevens and Coupe 1978). These studies demonstrate that mental maps are not continuous: people know, for example, the positions of places A, B and C in relation to each other and the positions of places α, β and γ also in relation to each other. However, they do not necessarily know the positions of places A and α or A and β in relation to each other. Rather, they are likely to know the relative position of a whole cluster of places ABC and αβγ (regions) to each other. As MacEachren (1992: 249) summarised it, "People seem to know the relative position of places within a region and relative positions of regions, but not the position of locations within different regions." This means that to travel from place A to place α, people take a course to the cluster (region) αβγ, rather than to a particular place α within it. It is only after their arrival in this region that they can find the particular place α. The size of regions thus encoded and their structure also seem to vary depending on the life history of an individual and, particularly, on the history and form of his or her traveling experience. This mental map hierarchy, therefore, may well also be an area of cultural variation.

To sum up the discussion thus far, human spatial cognition in any given moment of time involves a complex structure of representations that consists of a mental map divided into several regions and a set of routes between places within each of these regions as well as outside the regions. The term "cognitive map" is proposed here to refer to this structure. In this sense, the term "cognitive map" conveys not a particular form of representation (in contrast to a mental map), but rather a structured system of representations of different forms that is used by a subject to guide his or her spatial behaviour. One can expect considerable differences in the size and hierarchical structure of mental maps and in the frequency of use of mental maps versus route knowledge between different individuals of the same group, as well as between different peoples. This expectation is based on the finding that all these parameters appear to depend on the sex, travel experience, and life histories of individuals, which, in turn, depend on an individual's way of life (e.g., settled versus nomadic).

Thus far our discussion has been limited to the structure of cognitive maps as spatial representations. Now it is time to say a few words about the forms of their content. According to Golledge (1999), cognitive maps can include three types of elements, each of which has distinct cognitive properties (note: these types have little to do with the types of environmental objects being represented). The first type consists of simple zero-dimensional objects or places. These elements can be incorporated into a cognitive map by means of both route knowledge (as a point on a route) and a mental map (representation of their geometric position in relation to other elements). They are used to represent environmental entities without representing their topology or internal structure. A great variety of environmental objects can be represented by these elements, such as hills, mountains, rocks, trees, forests, lakes, and buildings. The second type includes one-dimensional elements usually called lines (Golledge 1999). These elements can also be incorporated both as route knowledge (as a route) and as part of a mental map, and can represent elements of an object's topology, particularly its length. Such environmental objects as roads, paths, rivers, ravines, and power lines can be represented on a cognitive map as one-dimensional elements. Finally, the third type consists of two-dimensional surfaces (2-D-S elements) or "areas" (Golledge 1999). These elements are capable of representing the spatiality of environmental objects by taking into account their topology and internal structure. They can be effectively used to represent such environmental entities as forests, lakes, fields, and neighbourhoods.[27] They are the only elements which, it would appear, cannot be incorporated into a cognitive map by means of route knowledge and, therefore, demand a mental map for their representation. It should be noted, however, that exactly how these elements are encoded on a mental map is still unclear. Golledge (1999) hypothesises that subjects encode them by encoding their spatial borders as closed curved lines (one-dimensional objects) on a mental map, but this approach is difficult to justify, especially when one takes into account the fact that these elements can have rather vague borders (Allen 1999).

The objects represented by these three types of elements vary significantly between groups and individuals when their cognitive maps of the same environment are compared (Appleyard 1970; Golledge and Spector 1978). It is important to note that no

27 Many of these elements can also be represented as zero-dimensional places rather than as 2-D-Ss. However, representing them as 2-D-S elements allows a person to know their borders, form, and structure (paths inside a forest, houses inside a neighbourhood, etc.).

cognitive map can include all objects even in a tightly circumscribed part of the environment. However, detailed a cognitive map one had of, for example, one's native town, it would hardly represent as independent elements all the houses, let alone all the trees in its park or all the stones on its streets (Downs and Stea 1973). Which objects are represented on a cognitive map depends on a number of factors at the group and individual levels. There are certain environmental elements that are likely to be represented on the cognitive maps of most if not all people who have knowledge of a certain environment or territory. For example, all people who have been living in a European city for more than a brief period of time are likely to represent on their cognitive maps its central square, town hall, most cultural, commercial, and recreational facilities, basic landmarks (like Big Ben in London or the Eiffel tower in Paris), and so forth. On the other hand, such an object as, say, the Institute of Anthropology (if the town has one) is likely to be represented on the cognitive maps of a much more restricted group of people. An individual's house is likely to be represented on the cognitive maps of only those people who are somehow related to that person. Finally, the cognitive map of a certain person could include such elements as "the place where I was hit by a car," or "the place where I kissed my future wife for the first time." These elements are likely to be different for every person.

The completely idiosyncratic elements (such as "the place where I kissed my future wife for the first time") give a clue to a very important fact about cognitive maps that should be highlighted. In contrast to the elements representing, say, the Eiffel tower or a town hall, these idiosyncratic elements, which are probably quite numerous on any cognitive map, may not be associated with any environmental object at all. "The place where I kissed my future wife for the first time" could be situated in a barren field, and the only thing that would make it different and account for its representation would be the particular experience related to it. This suggests that what the elements of a cognitive map actually represent are not the *physical objects* per se, but rather the *spatial aspects of knowledge and experience*, which are often (but not necessarily) related to particular physical objects (cf. Downs and Stea 1973). This fact, which remains in complete agreement with the primary function of a cognitive map as an organiser of experience, becomes even more evident if we take a look at the 2-D-S elements. As established by Lynch (1960), most of these elements do not represent physical objects. On the cognitive maps of cities inhabited by modern urban dwellers, these elements (often overlapping) mostly represent entities, such as parts of a city, neighbourhoods, quarters with buildings of a particular architectural style, areas with high traffic at certain hours, areas with high crime rates, areas having intensive nightlife, and areas where friends can be met after work. This does not mean that the 2-D-S elements cannot represent individual physical objects such as forests or lakes. However, when they do so, it is often in relation to an accumulation of experience and information concerning the internal structure of these objects. For example, a lake is likely to be encoded as a zero-dimensional place in the mental map of most of the people who use its shore for picnics. However, a fisherman who knows good fishing spots on this lake, its depths and hollows, and the landing places on its shore is likely to hold the representation of this lake as a 2-D-S element. Such representation seems to be absolutely necessary to encode his specific knowledge.[28]

The role of 2-D-S elements in organising environmental knowledge spatially is indeed overwhelming. These elements make it possible to structure this knowledge by sorting out the representations of local experiences and identifying, spatially, groups of places with certain properties. Furthermore, they allow inferences regarding the likely properties of other places never before visited and, therefore, guide spatial activity in important ways. For example, a person who has mapped the 2-D-S element "the medieval part of the city" onto his or her cognitive map has a good idea where to search for previously unseen or unnoticed medieval buildings. The information that a new restaurant has just been opened can influence one's plans for the evening very differently if one figures out that it is situated in an area with a high crime rate or if one sees that it is in the best part of the city. Of course, these inferences can be unreliable: the person searching for medieval buildings may discover that the previously unseen areas inside the medieval part of the city contain only modern buildings. In the same way it could be found that the new restaurant situated in the best part of the city is not very good. These occurrences can transform both the elements (the borders of the medieval city on the cognitive map) and the information represented by them (the belief that a certain 2-D-S element represents the best part of the city). However, this would not discourage further inference on the basis of our perceived structure of the city represented by means of 2-D-S elements, since there is no other reliable means of inference upon which to rely. Taking into account this role of 2-D-S representations in guiding practical activity and inferences, it can be expected that these elements are particularly sensitive to the demands and challenges of culturally specific practices and, therefore, vary tremendously between cultures and groups.

The rather detailed account of structure and content of cognitive maps given above allows a more targeted search for the ways the spatial cognition of reindeer herders can be adapted to the tasks of reindeer herding practice. Indeed, this adaptation can be expected to involve those elements of spatial cognition where culture can play variations on cognitive universals. Three such elements have been identified:

28 It could be argued that lines (one-dimensional elements) integrated into a cognitive map by route knowledge should have the strongest association with the physical objects. Indeed, it is the visual images and perspectives (vistas) of these objects that make up these elements' internal representation. However, even these elements represent the physical objects along the route *as well as* the knowledge and experience related to them. As Downs and Stea (1973:10) point out, even a person "who knows only one route, knows more about this route than simply the appropriate responses that should be made at certain choice points." Thus the knowledge of the route, say, from home to work includes, apart from the vistas and transitions, the knowledge of traffic intensity at different hours, of road conditions under different weather conditions, of the people one is likely to meet on the way, etc.

the relative role of route knowledge and mental maps in the overall cognitive map; the structure of the mental maps (their hierarchy, orientation, and ways of encoding structural information such as direction and distance); and the number and kind of content elements, particularly 2-D-S representations. The discussion of the spatial cognition of reindeer herders in the last section of this chapter as well as in the following two chapters will be focused on these three elements. Before proceeding to this discussion, however, some words still need to be said about methods of research on spatial cognition.

3.3 Methods of research on human spatial cognition "in the wild"

Studies on cognitive maps in psychology, geography and cognitive science are normally performed in laboratories and can involve orientation experiments in virtual environments with careful control of parameters. Many recent studies make use of neuroimaging techniques such as fMRI to observe directly the functioning of participants' brains in the course of these experiments. Of course, all these laboratory techniques would be of little use in our study on reindeer herding nomads. Fortunately, a long tradition of studying spatial cognition "in the wild" also exists. Field investigations were the main type of studies during the 1960s and 1970s, and a relatively effective methodology for such studies has been worked out.

Investigations "in the wild" are usually based on two methods: pointing experiments and sketch map analysis. The former method consists of testing a person's ability to point accurately to certain places currently out of sight. It is believed that such an ability is a good indicator that the person has an internal representation of the geometric position of the place in relation to him or herself, that is, that the place and the current position of the person are both represented by means of a mental map in a single region. In the words of Golledge, "pointing accuracy is usually a good indicator of whether or not layout understanding has been achieved" (Golledge 1999, 41). By asking a person to point to a large number of places from different locations it is possible, therefore, to clarify the size and regional structure of his or her mental map (Levinson 2003; Widlok 1997).

The sketch map analysis consists of analysing the way in which people represent the environment on handmade sketch maps. In studies based on this method, participants are normally asked to draw a sketch map of a certain area with which they are familiar (usually a town or city). The sketches can then be analysed with respect to their overall style, their content (what environmental objects are represented and how), the distortions in the locations of the elements on the map in comparison to their objective locations, and the details of the drawing process (especially of the order in which different elements of the sketch map are drawn) (Appleyard 1970; Craik 1973). It is believed that sketch map analysis can provide information about the structure and elements of the cognitive map of the sketch map author. For example, if a person draws a line (representing, for example, a street) first and then places the objects along this line in a certain order (the so-called route-based sketch map – Appleyard 1970), it is likely that he or she is communicating route knowledge. It was demonstrated, however, that the choice of elements to be represented on a sketch map, its style, and even the process of drawing greatly depend on the purpose for which the map is drawn. For example, if the map is drawn in order to explain a way to a certain place or places, or to recount a particular journey, the chances are that it will be built around a line (or lines) representing a route (or routes), and contain only those elements that can be used as landmarks for wayfinding or that are important to the story. This does not mean that the lines are indeed integrated into the author's cognitive map by means of route knowledge rather than inferred from a mental map, nor does it mean that all the elements of the cognitive map are represented (Spencer and Weetman 1981). Nevertheless, sketch map analysis is believed to be an effective way of studying cognitive maps of objectively existing environments if the conditions related to drawing the sketches are carefully controlled. The analysis of large numbers of sketch maps of the same territory produced by different people can be used to clarify group differences in cognitive maps (Montello et al. 1999; Moore 1976; Spencer and Weetman 1981). Similarly, the analysis of sketch maps produced by the same group of persons at different times has been used to study cognitive mapping processes (Delvin 1976). Recent field studies on spatial cognition have also made use of the analysis of spatial communication, particularly of the manner in which people explain the way to different places to other people or to the researcher. These explanations, which normally include verbal descriptions, references to certain landmarks, and gestures, are believed to convey important information about the structure and elements of that person's cognitive map.

As can be seen from the account of these methods, the studies of human spatial cognition in psychology, geography, and cognitive sciences have been heavily dominated by the analysis of external representations of this cognition (called "spatial products") in the form of gestures, graphical images (sketch maps), and the performance of different spatial tasks. In almost all cases, these spatial products are created at the special request of the researchers during the course of experiments and analysed by means of quantitative methods. Although these methods have been rather effective in providing basic knowledge of human spatial cognition, they clearly have their limitations – a fact readily acknowledged by the researchers themselves. They give no information on how cognitive maps are actually used in everyday life to guide people's behaviour, nor do they shed light on what the exact factors are that account for individual and group differences in cognitive maps. In order to understand this, it would appear that qualitative methods of participant observation and interviewing need to be applied. This means that anthropology has its place and role in the interdisciplinary research described above, and its integration into the family of disciplines studying human spatial cognition should be fruitful.

For these reasons the research methods applied in our studies on spatial cognition among reindeer herders included pointing experiments and sketch map analysis, as well as standard anthropological methods of participant observation and interviewing. The pointing experiments, the sketch maps, and the interviews were used to ascertain the size and structure of the informants' cognitive maps and the territories they cover, as well as to gain insights into their content. Participant observation was used mainly to understand how cognitive maps are employed for wayfinding and organising other spatial activities. It included traveling with informants for various purposes (participating in herding procedures, making trips with reindeer herders to villages, going hunting and fishing with settled informants, etc.). During these trips, the informants were asked how they chose a route, why exactly it was chosen, what the properties of the places we went through were and how they were used, and so forth. The information obtained was later verified in interviews.

3.4 Cognitive maps of Komi and Nenets reindeer herders: General structure

It follows from the theoretical discussion presented thus far that the first aspect to consider in research on the spatial cognition of reindeer herding nomads is the general structure of their cognitive maps. This would include specifying the relative roles that the two basic forms of spatial representation – mental maps and route knowledge – play in them, as well as determining if reindeer herders somehow differ in this respect from other groups living in a similar environment. Once this first step is made, the particularities of the two forms of spatial representation and of the content of the reindeer herders' cognitive maps can be further discussed.

As was noted above, one important factor that can potentially affect the general structure of a cognitive map is language and, particularly, the frames of reference a given language routinely utilises for specifying relative positions of objects. From the viewpoint of our research question, it makes sense to consider this factor first because, it can be argued, a language's grammar is not directly related to the economic activities and the way of life of its speakers. Therefore, if the Komi and Nenets languages routinely employ different frames of reference and, particularly, if they employ frames of reference that are different from those employed by sedentary groups, then the particularities of the herders' cognitive maps can potentially be explained by these linguistic differences rather than by the specific cognitive tasks they face in the course of their everyday activities.[29] This, of course, could be detrimental to the general argument we wish to develop here.

29 Of course, one can still claim that the prevalence of specific linguistic frames of reference in a given language can be caused or fostered by particular practical needs of people who speak it or spoke it in the past. This claim, although interesting, demands a great deal of evidence that is difficult to obtain. For this reason, it is not explored here.

Both the Komi and the Nenets languages belong to different branches (Finno-Ugric and Samoyedic respectively) of the same (Uralic) linguistic family. In both languages, spatial relations between objects are expressed by means of localising cases of nouns (encoded in noun suffixes) and postpositions. The latter also acquire the localising cases' suffixes. The localising cases of nouns express mostly the relations of inclusiveness (something being inside something) and contact between objects. For example, in Komi, "inside a nomadic tent" is *chomiyn*, derived from *chom* (nomadic tent) and *-yn* (inessive suffix). In Nenets, the same phrase would be rendered by the word *makona*, from *ma'* (nomadic tent) and *-hana* (inessive suffix, rules of phonetic assimilation apply). The spatial relations other than those of inclusiveness and contact are designated mostly by postpositions. Thus, "behind a nomadic tent" in Komi would be *chom saiyn*, from *chom* (nomadic tent) and *sai* (space behind), plus the inessive suffix *–yn*. In Nenets, it would be *ma' pun'ana*, from *ma'* (nomadic tent), *pun'* (space behind), and *-hana* (inessive suffix). Although some of the postpositions in both languages semantically express spatial relations in the absolute frame of reference (for example, Komi has *chom katyn*: "upstream from a nomadic tent"), their use seem to be restricted to a limited number of situations. For example, the postpositions *kat* (space upstream) and *kyvt* (space downstream) in Komi are used only when speaking about objects situated on a bank of a river or stream. Most spatial expressions involving postpositions in both languages clearly use the relative and intrinsic frames of reference.

In the Komi language, two verbs describing movement in the absolute frame of reference were observed: *kainy* ("to move toward the nearest watershed divide") and *löchchyny* ("to move away from the nearest watershed divide"). These verbs are frequently used to describe movement on both macro- and micro-scales, even when no streams or rivers are visible. Thus the sentence, *Löchchy rösshaas da pes kayöd!* ("Go away from the divide to the bush stand and bring some firewood back up here toward the divide!"), can be used only when the speakers are situated between the watershed divide and the bush stand. However, these verbs co-exist with equally frequently used verbs describing movement in the intrinsic frame of reference. In the Nenets language only "intrinsic" verbs were recorded (although we must acknowledge that our familiarity with this language is much more limited than with that of Komi). Nevertheless, it can be supposed that both peoples rely predominantly on intrinsic and relative frames of reference in their speech. Furthermore, they do not differ in this respect from speakers of Russian, who comprise the majority of the settled population in both regions. This provides some assurance, albeit qualified, that the differences in the structure of cognitive maps between the two reindeer herding groups, as well as between the reindeer herders and the settled population, are not related to the languages they speak.

In the course of ethnographic interviews both Komi and Nenets reindeer herders frequently stressed the importance of "knowing the land," and even the younger herders had detailed knowledge of their territory. Six male Komi reindeer herders from two reindeer herding brigades sharing the same migratory path (*vörga*)[30] were interviewed while performing their duties near the herd of productive animals (*kör*)[31] and asked to describe the land around their current position. All the herders were able to produce a detailed description by listing rivers, streams, lakes, ponds, bogs, *tabey* (swampy areas),[32] willow stands, hills and hill ranges (*musur*), and areas characterised by a specific type of vegetation (e.g., *pacha* – an area covered with dwarf birch; *yarey* – an area with a lot of lichen; *nyar* – an area without vegetation). The description was almost always accompanied by pointing gestures meant to identify the direction in which these landscape features were situated, and sometimes their approximate distances from any given point would also be indicated. The informants also often provided details of the relief of the terrain they described. A peculiar gesture was uniformly used for this: an informant would move his finger along the line of the horizon following the course of the river situated behind it; while doing so, he would bring his finger closer in or move it further from his body to describe the zigzags of the riverbed. Most of the features described by the informants were out of the line of view from the place where the description was given. This behaviour strongly signifies that the Komi informants had a very clear idea about how the various features they listed were situated in relation to each other at any given moment in time. In other words, they possessed a mental map (survey knowledge) of the environment.

Further investigation into the cognitive maps of Komi reindeer herders was performed by means of sketch-map drawing tasks and pointing experiments. Twelve male reindeer herders were involved in this study.[33] The research revealed the existence of clear spatial limits to the territories they represented in the form of mental maps. Thus all the informants could provide very detailed sketch maps of the position of various landscape features situated on both sides of their migration path (*vörga*) at a distance of up to around 15 km away. They could also point to these features with rather amasing accuracy (the mean error being 8 degrees, with a standard deviation of 1.2 degrees). However, the territory beyond this limit was accounted for in much less detail. It was noted, for example, that while speaking about territory close to their path, the herders could tell about and point to areas with specific vegetation, bogs and *tabey*. In more distant territory, only objects distinctly standing out from their environment and visible from far away – landmarks such as large rivers, lakes, high

30 Brigades 2 and 3 of the Bolshaia Inta reindeer herding enterprise (see chapter 1 for details). The interviews took place in summer 2003.

31 See chapter 2 for details of the pasturing technology of the Komi.

32 A Komi term for a common natural phenomenon caused by the thawing of permafrost, which renders the ground unstable so that it moves and shakes when walked upon (not unlike jelly). In English this is generally rendered as "thermokarst." For a more detailed discussion of the landscape terminology, see chapter 5.

33 Of the twelve, eight herders were tested in reindeer herding camps in the tundra, while four herders were tested in the city of Inta. The research took place in winter 2008-2009.

mountains, and villages – were named and pointed to with accuracy (with an error ranging from 7 to 15 degrees).[34] The size of the territory these "semi-detailed" mental maps covered (i.e., from 20 to about 80 km on either side of the migration route) seemed to vary greatly from one reindeer herder to another. The territory beyond this distance seemed to be almost completely unknown to the herders and was referred to as "unknown land" (*tödtöm mu*). It seems, therefore, that the mental maps of Komi reindeer herders cover a long and relatively narrow strips of land situated on both sides of their migration routes, and that these maps have different levels of detail: they are elaborate in areas adjacent to the migration route (i.e., the core map), but gradually become vague and not sufficiently detailed to easily pasture reindeer the further one moves away from it (i.e., the periphery map).

As one would expect, the areas covered by the "core" maps in the case of Komi reindeer herders generally coincide with the territory they normally use for maneuvering their herds in the course of their day-to-day pasturing activities.[35] On the other hand, the "periphery" mental maps of the herders seem to be insufficiently detailed for successful reindeer pasturing. Thus, when asked whether they could pasture reindeer outside the normal pasturing corridor of their brigade, several informants replied that they could not because they did not know the land well enough. However, the "periphery" mental maps still seem to be detailed enough for wayfinding. Thus, all the informants, including those who claimed that they would not be able to pasture a reindeer herd outside their pasturing corridor due to inadequate knowledge of the land, insisted that they could easily find their way to any point in an area that was far greater than the one covered by their core maps. In the course of our fieldwork, we did indeed observe on a number of occasions how reindeer herders performed trips to the camps of other brigades situated several dozens of kilometers from their own. In the majority of these cases, the only information they had before starting the trip was the name of the place in which the other camp was situated, which had been communicated to them by radio. Many herders also stated that although the land beyond a certain distance from their migration path was unknown to them and they could not point to any location there, they still could find their way to certain points in it (most notably villages and towns) because they "knew the route there." This suggests that certain points in the territory outside that covered by the mental maps could be integrated into the cognitive maps of the herders by means of route knowledge.

34 It should be noted that informants often refused to point to a feature, saying that they were not able to do this with enough accuracy. This response was taken as an indication that this feature was not a part of his mental map. When the informants did agree to point to the feature, however, their error in the great majority of cases did not exceed 15 degrees. In fact, there were only three exceptions, and all of them could be explained by the informants' misunderstanding of the specific place the researcher wanted them to point to.

35 As was described in chapter 2, the land a Komi reindeer herding brigade can use for pasturing its herd represents a corridor between its own migration path (*vörga*) and that of a neighbouring brigade to the east or west of it. This corridor is normally 10 to 20 km wide.

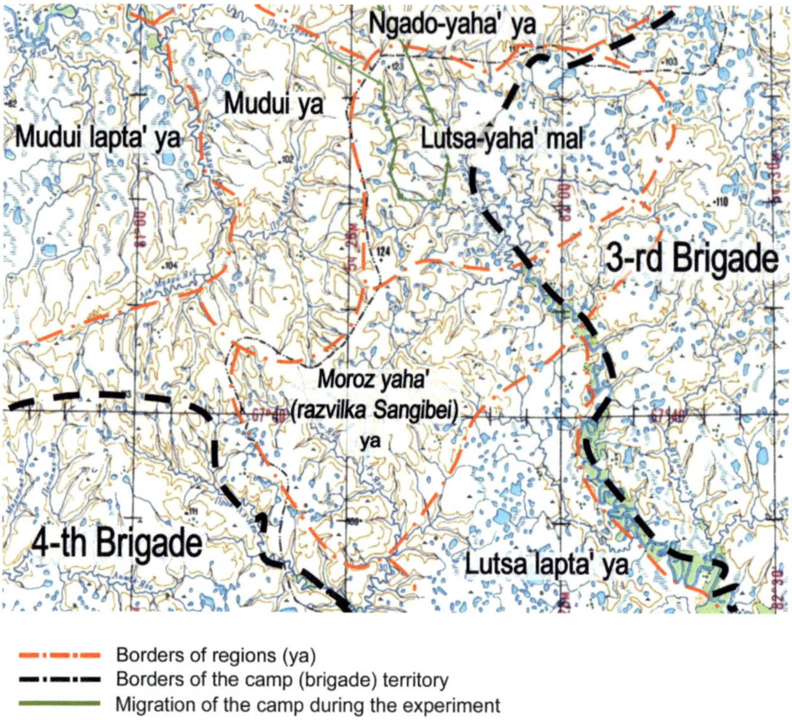

Fig 3.1 Nenets' conceptualisation of land as a set of regions (*ya*), the area of upper Lutsa-yaha, Gydan Peninsula.

The mental maps of the Taz Nenets reindeer herders, who do not have migration routes, appear to be different. Like the Komi, the Nenets herders, at any time, could list a great number of landscape features that surrounded their territory, including rivers, bogs, forested patches, willow stands, lakes, hills, and ponds that surrounded their territory. These features were pointed to and, as with the Komi, some informants provided an approximate distance to them. The sketch-map tasks and pointing experiments performed with twenty-eight male herders aged between fourteen and forty-eight[36] revealed that most of them (twenty-one) could name and point with high accuracy (with an error of less than 15 degrees) to landscape features situated as far as 25–30 km away in any direction from their location. Beyond this distance their knowledge of the terrain became much less detailed, similar to the Komi, although significant differences among the participants could be observed. Their detailed mental maps, therefore, seemed to cover a circular tract of land with a radius of about 25–30 km.

36 The herders belonged to three reindeer-herding brigades. The investigations among these groups took place in summer 2006, summer 2009, and winter 2009–2010 respectively.

The most important difference between the mental maps of the Komi and the Taz Nenets reindeer herders, however, is that the mental maps of the Taz Nenets seem to have a much more complicated structure. It was revealed that the Nenets conceptualise their territory as a set of regions (called *ya* in Nenets)[37] that mostly coincide with watersheds, each of which has its own name and relatively clear borders that follow the ridges separating two adjacent drainage basins (see fig. 3.1.). These regions are relatively small in size and cover a territory of approximately 25 x 25 km. A reindeer herding group usually visits three to four of these regions during the course of its annual migrations. Most male Nenets herders could easily point to any place inside the region (*ya*) where they were currently situated, but most of them had significant problems pointing to places situated in neighbouring regions when they were asked to do so. For example, during the entire period that one of the groups that we investigated was situated in the region called *Lutsa yaha' mal* ("the land in the upper part of the river Lutsa"), nine of eleven of its male members who were more than fourteen years of age could point to all the places in this region with a high degree of precision (the error being less than 15 degrees).[38] However, seven of them had difficulty pointing to the old drilling tower near the N'ado (*Ngado*) river that was situated in the region called *Ngado-yakha ya* ("the land of the Ngado river") just to the north of the *Lutsa yaha' mal.* When asked to point to the tower, these herders simply pointed in a northerly direction (which produced an error of approximately 21 degrees) and said that the tower was in *Ngado-yakha ya*. Once the herders crossed the ridge between the watersheds of the Lutsa and N'ado rivers (and, therefore, entered the *Ngado-yakha ya*), eight of the eleven informants became able to point to the tower with much greater precision (the error being less than 15 degrees) even though the distance to it had actually increased. On the other hand, eight informants could no longer point with an equally small error to the places near the river Lutsa (something that six of the eight were able to do previously), despite the fact that the distance to some of those places was less than it was to those objects in the N'ado River region that five of the eight could now point to from their new location. They accounted for this by saying that the camp was in "another land" (*ya*) now. These results indicate that their ability to point to a location depended on the region in which they were situated rather than on the actual distance to those objects. This effectively rules out the possibility that their willingness or refusal to point to the locations reflected their awareness of limited precision of pointing gestures in the case of distant objects or their interpretation of

37 The word *ya* has many meanings for the Nenets, the most general of which are "land," "ground," and "earth" (as opposed to *num* – the sky). In the few Nenets radio broadcasts and the Nenets-language newspaper published in the Yamal-Nenets autonomous area, this word is also used in the sense of state or country (e.g., Russia is referred to as *Lutsa' Ya*, which literally means "the land of Russians").

38 The investigation reported took place in summer 2006, in the camp of Brigade 1 of the Tazovskii reindeer herding enterprise.

the task (i.e., detailed spatial answers only make navigational sense for a limited area). Rather the data shows that in contrast to the Komi, the mental maps of the Nenets have a hierarchical structure as described in section 2 of this chapter.

The following description provided by one of the informants explains both the organisation of Nenets mental maps and their hierarchical structure more precisely:

> The land between the Vyder-yakha and Limb'a-yakha [names of rivers] is the one I know best. When I was young, I used to "live" [meaning "to migrate"] there for a long time. There I can point every tussock out to you. There, I can point to everything and from everywhere. I also know the land down from the village [of Tazovsky], between Vesako-yakha and Koptan [names of rivers]. We "lived" there for a long time as well and I know every tussock there also. But I cannot show the direction to them from here. If we were there I could, but not from here. (Sergei Khudi, age 42, interviewed in Taz tundra, November 2005)

This quotation – one of several comments of this sort obtained during interviews – reveals that the Nenets' "knowledge of land" is uneven. Nenets herders seem to possess a more detailed knowledge of some regions than others due to their personal life histories. The more detailed the knowledge is, the more detailed the mental map of the region appears to be. The important point, however, is that the detailed mental maps of the regions are independent of each other – the informants insisted that once inside the region they can point to "every tussock" from any given point. However, they were unable to tell how most of the places inside the region were situated in relation to their position outside of it. This also falls in line with the psychologists' and geographers' argument about the hierarchical structure of mental maps.

The cognitive maps of the territory outside these regions seem to be much less detailed. The accounts of this territory provided by the Nenets herders usually included only basic landmarks, among which rivers were the most common. Indeed, when asked to describe the territory outside their "pasturing region," the herders typically listed rivers, pointing to the direction in which they were situated and specifying the direction of their current. Although the general direction of these pointing gestures was almost always correct, it is difficult to interpret them in term of the cognitive map structure. Indeed, since the herders pointed to the river in general rather than to a particular point on it, and since the rivers in western Siberia tend to be rather long, it is difficult to say if the herders know the detailed position of the riverbed vis-à-vis specific points in the landscape (in the form of a mental map) or just a general direction in which the river stream could be reliably encountered. The latter seems to be more likely, since the herders usually refused to point to any specific locations along the river stream (such as the mouths of its tributaries) unless those locations were situated in the region (*ya*) that the herders were currently in or, sometimes, in one of the neighbouring regions. Whatever their structure might be, the area covered by these "hydrosystem" cognitive maps was rather huge, albeit with significant variation in size among informants. Informants admitted that the area known to them varied depending on the age and experience of a herder: older herders generally knew a larger territory than younger ones. However, even the "hydrosystem" cognitive maps of younger herders seemed to cover surprisingly vast areas. For example, one sixteen-year-old reindeer herder managed to list the rivers to the east of his current location up to the Yenisei (about 300 km away). He stated that the Yenisei river had several large interconnected branches and was located in a huge valley. He also described the river's eastern tributaries and showed by gesture the directions in which the Yenisei and Kheta rivers flowed before they met.

It seems probable that rivers represent the main axes of Nenets spatial representations. The position of every location is made with reference to a particular section of a river (e.g., lower, middle, and upper). The locations of regions for which the Nenets had detailed mental maps were typically described as being between such and such rivers. Furthermore, any movement outside the camp was referred to as descending or ascending with reference to a river running nearby, despite being some distance away and often out of sight. Therefore, the cognitive maps of the Taz Nenets reindeer herders apparently have two hierarchical levels. The upper level consists of the "hydrosystem" map, which means that regions are known in relation to the locations of rivers and directions toward them. The lower level consists of detailed mental maps of some of these regions. It seems that the Nenets are quite aware of the existence of these two levels of "knowing the territory," as the following interview excerpt indicates:

> Nenets herder: It would be nice to go and live on Lutsa-yakha, it is a good place. There is plenty of lichen, a lot of fish, many animals, and no Russians there. No Russians on Lutsa-yakha.[39] However, we cannot go and live there alone; we would need someone with us who knew the land there.
> K.I.: You do not know how to get there?
> Nenets herder: Why, yes we do. Of course, we do not know all the hills, bogs, and lakes! But we can still find our way there. We could travel from one river to another and eventually get there. Everybody knows rivers, right? But to live in a place one has to know the land and not just the rivers. You need to know all bogs, lichen places, where to pasture reindeer. To know the land, we need an old man (*vesako*) to go there. (Interview with Taz Nenets herder conducted in the Taz tundra, November 2005)[40]

This interview shows the conceptual difference the Taz Nenets made between "knowing the rivers" and "knowing the land." "Knowing the rivers" seems to refer to having the upper level of the cognitive map, which, the informant believed, everyone has. "Knowing the land" refers to the detailed lower-level mental maps. Apparently, as described in the interview, nobody in the camp had these detailed maps and this prevented this particular group of Nenets from moving to the territory.

The research performed in the Taz tundra region included a comparison of the cognitive maps of the reindeer herders with those of semi-settled Nenets fishermen,

---

39 *Luts'a-yaha* literally means "Russian river."
40 This particular herder chose to remain anonymous.

as well as of representatives of different professional groups from the mostly Russian-speaking settled population. Two such groups were examined: helicopter pilots working for the local aviation company (UtAir) and gas drillers working for the Gazprom company. Five male helicopter pilots aged between thirty and forty-five, five gas drillers, and seven semi-nomadic fishermen participated in pointing experiments in the town of Tazovskii (the settled groups) and in the camp of Nenets fishermen in Habdu-ya (situated in the Taz river floodplain about 30 km upstream from the town of Tazovskii).[41] In addition, two representatives of each group were asked to draw a sketch map of a territory adjacent to the town of Tazovskii.

This investigation revealed that the helicopter pilots were able to name and correctly point to locations (with an error of less than 17 degrees) situated at distance of up to 100 kilometers from the town of Tazovskii. This indicates that, as far as the distance to the location is concerned, the performance of helicopter pilots far exceeded that of the reindeer herders. On the other hand, the number of locations in the tundra that the helicopter pilots were able to name and point to was much smaller than that of the reindeer herders. These locations tended to represent natural and built objects that were the most prominent in the surroundings, such as larger lakes, prominent hills, forest "islands," geodesic and gas-drilling towers, settlements, isolated buildings, and landing strips. The pilots could neither name nor point to areas with specific vegetation, smaller hills or lakes, bogs, or other natural features that characterised the reindeer herders' accounts of the territory. In short, although the cognitive maps of the helicopter pilots seemed to cover rather vast expanses of territory and their mental maps were, it seems, much larger than those of the reindeer herders, they also seemed to be much less detailed in terms of the number of objects included. Unfortunately, little more can be said about the structure of the pilots' cognitive maps on the basis of the material collected.

Although the semi-nomadic fishermen could list a number of localities in the territory that surrounded their fishing camp, particularly those situated inside the Taz river floodplain, they could not point to most of them. In fact, only two of seven fishermen could point to more than one-third of the localities they listed with an error of less than 17 degrees. These localities were situated up to 40 km from the camp. Most fishermen, however, said that they were not able to point to even to those localities that were situated less than 10 km from the camp (3 km was the minimal distance). The reason might be that their cognitive maps are heavily dependent on route knowledge as opposed to a mental map. Indeed, all of them could report a number of interconnected rivers, as well as explain how to reach certain places by travelling from one river to another despite being unable to point to these places. Finally, the smallest cognitive maps seem to be those of the drillers, who rarely leave the drilling installations during their stays in the tundra and, if they do, they usually stay within sight of

41 The experiments with the helicopter pilots and the gas drillers were conducted in spring 2007. The experiment with the fishermen was conducted in summer 2009.

the drilling tower. Among the five drillers who participated, only one could list more than ten localities in the tundra, and none of them could accurately point to any of the localities they mentioned.

These comparative results suggest essential differences between the cognitive maps of the reindeer herders and those of other groups populating the tundra. The obvious explanation is that these differences are related to the everyday practices of these groups. This will be investigated further in the following chapter.

All information regarding the spatial representations reviewed so far was obtained from males of fourteen years of age or older. Unfortunately, little information could be collected on the methods of spatial orientation used by either female informants or other age groups, but the empirical studies described in the previous section indicate that these could be quite different. Indeed, the difference in the methods and capacities of spatial orientation between children and adults, as well as between the sexes, was mentioned on many occasions by our informants. For example, some Komi reindeer herders made comments that support the theory that children rely mostly on route-based orientation:

> Children remember the way better than grown-ups, do they not? I could forget the way, but they would not. They do not know the land, they do not know where different places are, but if they were shown a route they would remember it and could always come back. (Interview with Vasiliy Yangasov, 36 years old, Bol'shezemel'skaya Tundra, August 2003)

Both the Komi and the Nenets have a common belief that women are less apt than men as far as wayfinding is concerned. Women themselves especially stress this view. For example, two young Komi women in the eastern part of the Bol'shezemel'skaya tundra mentioned that they had never been to the sea coast (situated about 15 km north of the northernmost point of their migration route) because the men had never taken them there. When asked if they could not go there by themselves, they answered, "We would probably end up in Vorkuta if we tried."[42] Similarly, Nenets women stated that they would never go anywhere without men because they could not orient themselves and, therefore, would get lost. Furthermore, both the Komi and the Nenets females gave much poorer descriptions of their territory, listed fewer features and almost never pointed in the correct direction.

It would definitely be misleading, however, to rely on this evidence to support the hypothesis that women have worse spatial orientation abilities than men. As was pointed out by Habeck (2005), among Komi adults, women are much more limited in their mobility than men. In fact, their mobility is restricted to driving a caravan of sledges along a migration route when movements take place (Habeck 2005:36–37), and is related to the traditional division of labor between the sexes, whereby women for the most part carry out domestic duties that keep them in and around the camp.

42 Vorkuta is a city in the Bol'shezemel'skaya tundra situated about 300 km away.

Habeck noted that younger girls who had not yet assumed the full responsibilities of tent-workers enjoyed a much higher degree of mobility than older women. This statement is supported by our observations: the mobility of young girls was observed to be quite high and essentially similar to that of boys of the same age, and no difference in their spatial performance was noted. The comment of the older reindeer herder on the spatial abilities of children quoted above referred to both boys and girls. Therefore, the difference in the wayfinding abilities between older men and women can probably be explained by their different degrees of mobility rather than by any differences between the sexes per se. The same is true for Nenets women, whose mobility is similarly (if not more)[43] restricted. On the other hand, since the Nenets do not follow established migration routes, their women probably become familiar with a larger territory during the course of their seasonal movements. This could explain their greater ability to describe their territory. These findings appear to support some psychological and geographical studies (e.g., Kitchin 1996) that argue that cognitive map knowledge and abilities among the two sexes are equivalent, and that any differences are likely to be socially and culturally produced, greatly influenced by gender roles and reinforced by stereotypes.

To sum up, the data on spatial cognition of Komi and Nenets reindeer herders presented in this chapter demonstrates that the cognitive maps of these two groups differ significantly from each other, as well as from the cognitive maps of other groups, in terms of both structure and form. Factors related to everyday practices, such as the flexibility of movement between herding groups (linear as opposite to circular), seem to be responsible for these differences. The Komi, who use established migration paths and pasture reindeer along them, were found to have detailed and continuous mental maps of these areas. In contrast, the Nenets, who do not have established migration paths and often change the territory of their seasonal migrations, have detailed mental maps of only relatively small and discrete regions. The Nenets do, however, have an additional level of cognitive spatial representation consisting of only basic landscape features (mostly rivers), which can be spatially related to one another. This finding confirms the idea of a hierarchy of mental maps and its dependence on a way of life, as proposed by geographers and psychologists.

The following chapter delves more deeply into the structure of these mental representations and their use in everyday practical activities.

43 Among Taz Nenets, women should not even touch (let alone sit on) male transport sledges, and the big, heavy sledges for the women, are unsuitable for traveling anywhere other than on established paths, particularly in winter, as they sink into the snow.

4 SPATIAL COGNITION AND SPATIAL NAVIGATION: A COMPARISON BETWEEN REINDEER HERDING NOMADS AND HELICOPTER PILOTS

4.1 Spatial cognition and navigation: Preliminary remarks

To continue the inquiry into the spatial cognition of the two groups of reindeer herding nomads and to understand how culture "plays variations" on it, let us now analyse how they use their cognitive maps to solve practical tasks. As was mentioned in the previous chapter, the range of everyday tasks that require reliance on the representation of space is likely to be extremely wide among reindeer herders. In this chapter, however, we shall focus on only one such task: the practice of spatial navigation (wayfinding), by which we mean the process of finding a way between particular locations in space. This choice is justified because, first of all, spatial navigation is usually considered an example *par excellence* of putting spatial cognition into practice, and an absolute majority of "classical" studies on spatial cognition (e.g. Allen 2004; Downs and Stea 1973, 1977; Golledge 1999; Siegel and White 1975; Tolman 1948 to name just a few) discuss their subject in relation to this practice. Furthermore, the specific practices of spatial navigation among representatives of particular cultures have often been the object of anthropological research (e.g., Aporta 2009; Aporta and Higgs 2005; Gell 1985; Gladwin 1970; Hutchins 1995; Ingold 2000; Lewis 1972, 1976; Nelson 1969; Porteus 1937; Widlok 1997). This means that spatial navigation is relatively well studied in comparison to other applications of spatial cognition, and that our study will contribute to an already established tradition of cross-cultural research. Secondly, wayfinding tasks are understandably among the most frequent and pervasive in the everyday lives of nomads and, therefore, the way such tasks are handled by reindeer herders is more accessible to field studies. Finally, and probably most importantly, spatial navigation, in contrast to more specific practices related to reindeer pasturing that also most likely involve spatial cognition (see chapter 3), represents an important part of everyday life for groups other than reindeer herders. This opens up the possibility of comparing the wayfinding practices of nomads with those of settled groups living in the same area, thereby providing access to the cultural variation in spatial cognition more directly.

The process of spatial navigation or wayfinding[44] can be generally understood as taking and keeping the course toward a desired location. As was mentioned in the previous chapter, both procedures can be modeled relatively straightforwardly in cases where route-based knowledge is involved in performing this cognitive task: the perceived visual image or vista of the environment (visual stimulus) is compared with the stored temporarily arranged images or vistas specifying the route to the location. The course-taking (response) is made so as to ensure the correspondence between these images. In practice, however, wayfinding by route knowledge should have mechanisms to adjust for a possible mistake in comparing represented and perceived images (misrecognising the place). In cases where navigation is based on a mental map, the additional procedures of updating one's position in relation to encoded locations (position fixing) and inferring the direction of travel to the target (course taking) have to be performed. These two procedures are sometimes jointly referred to as the "map lookup process" (Easton and Sholl 1995; Wang 2012). Only after one's position is fixed can the position of the desired location in relation to this fixed position be recalled using the mental map, which can allow inferring and taking a course toward it. During travel, this course (heading) should be maintained either by using internal mechanisms sensitive to a change in heading (which are, however, little developed among humans – see Loomis et al. 1999) or by employing some other method of course keeping. Again, there has to be a mechanism to account for possible mistakes in performing all these procedures. It is reasonable to suppose that there can be rather different ways of performing all these procedures (i.e., position fixing, course taking, course keeping, accounting for mistakes). This can give rise to diverse practices of navigation that can vary in their overall effectiveness as well as, importantly, in the kinds and patterns of mistakes they are potentially prone to. Furthermore, these practices of navigation represent a potentially important area of cultural variation that is accessible by means of comparative research.

In this chapter, we present an attempt at such an inquiry by comparing the navigation practices of Nenets reindeer herders from the Taz tundra to that of "visual piloting," the navigation method used by the older generation of helicopter pilots working in this region. In the previous chapter it was noted that the pilots seem to have particularly large cognitive maps that far exceed those of reindeer herders. They also

44 Tim Ingold (2000) finds it useful to differentiate between spatial navigation and wayfinding: in his opinion, the term "navigation" should be reserved for instances of finding one's way by inference from a mental or artificial map, while the term "wayfinding" should be applied only in instances of stimulus-and-response route following. In our opinion, this differentiation makes little practical sense. Indeed, as was described in the previous chapter, any cognitive map is likely to be based on both mental map(s) and route knowledge. Correspondingly, many (if not most) instances of finding one's way can be based on inference from a mental (or artificial) map and route following, both of which can be applied either simultaneously or at different stages of traveling. Therefore, in this book we use the terms "navigation" and "wayfinding" interchangeably to refer to the process regardless of the cognitive mechanisms supporting it.

have extensive experience in spatial navigation, albeit under rather specific conditions. In our opinion, this makes the comparison particularly telling. For this comparison we rely mainly on data collected by means of participant observation and ethnographic interviews, paying particular attention to the patterns of mistakes and hardships experienced by the representatives of the two groups during navigation. Unfortunately, no comparable material on navigation practices has been collected either among the representatives of the second group of reindeer herding nomads in our study, the Izhma Komi, or among the semi-nomadic Nenets fishermen. Still, some available qualitative information on these groups is reported, where relevant, in the context of the comparison.

4.2 Wayfinding in visual helicopter piloting

The practice of navigation used in visual helicopter piloting is a good place to start the comparison because it is highly formalised and relatively unambiguous. In addition, the very special features of this practice make its effect on spatial cognition (and vice versa) especially visible. It is important to mention, however, that the practice of visual helicopter piloting is rapidly becoming obsolete with the introduction of satellite geopositioning systems (GPS), which started being used in the Taz tundra in the middle of the 1990s. Nowadays most helicopter pilots in the Taz tundra rely heavily upon GPS for navigation. The majority of them (in fact, all who started their careers as pilots at least eight years ago) are, however, still skillful in visual piloting, and many still use this practice in conjunction with GPS-based navigation.[45]

In order to understand the practice of visual helicopter piloting, some preliminary remarks on the social and technical organisation of a pilot's work are necessary. Helicopters have been the main means of transport (as far as the number of transported people and amount of cargo are concerned) in the Taz tundra, as well as in most parts of Siberia, since the late 1960s. The main type of helicopter used in all these areas has always been the famous MI-8, constructed in the late 1950s and put into mass production since 1963. Each MI-8 helicopter is normally operated by a team of three people: the first pilot–commander (*pervyi pilot–kommandir*), the second pilot–navigator (*vtoroi pilot–shturman*), and the on-board engineer (*bortinzhener*).[46] The first pilot is

45 Many older helicopter pilots in fact reported their preference for visual piloting over GPS-based navigation, which, in their opinion, makes the young pilots too dependent on the high-tech satellite equipment and unwilling to learn about the geographical environment. They also strongly believe that GPS systems are dependent on satellites that are controlled by the American military, which could switch it off at any time, thereby grounding all the flights if the methods of visual piloting were forgotten.

46 The on-board engineer does not have a pilot's license and therefore cannot properly be called a pilot. He cannot be promoted to pilot unless he completes training in a special flight school.

Mi-8 helicopter in the tundra. Photo: Mark Dwyer.

responsible for actually flying the helicopter, while the primary responsibility of the second pilot is navigation – he can be given engine control only for short periods of time during a safe period of flight to allow the commander to rest. The engineer is responsible for observing the engine equipment and ensuring its proper functioning during the flight. In accordance with a practice that emerged in Soviet times, a young pilot always starts as a second pilot responsible for navigation. His promotion to first pilot usually takes many years, as the general policy of the air companies (which also follow a tradition established long before the collapse of the Soviet Union) consists of forming crews on a long-term basis and does not support the transfer of people between crews. For this reason a second pilot usually becomes a commander only after the previous commander in his crew retires or leaves the company. An important effect of these policies is that the helicopter commanders always have a great deal of experience in navigation, which they effectively transfer to the second pilots.

The navigation equipment of the MI-8 helicopter, apart from the newly introduced GPS devices, consists of a magnetic compass, speedometer, two altimeters, and a radio-compass engine that can automatically determine the position of the helicopter by triangulation with stationary radio beacons. The essential precondition for such triangulation is that signals from at least two radio beacons are received concurrently. However, the region simply lacks a dense enough network of radio beacons to ensure the reception of two signals at once from an altitude of about 1 km (the usual altitude of the helicopter flight). In fact, only three radio beacons have been installed in the entire Tazovskii *raion* (and only one of them – in the town of Tazovskii – is in the southern part). These radio beacons are supposed to be used by planes that fly much higher and, therefore, can receive radio signals from greater distances. As far as helicopters are concerned, only one radio beacon is "visible" to them (i.e., the signal can be received) in most parts of the Taz tundra, while in the rest of the tundra no radio beacons are visible at all. This effectively renders the radio-compass useless except when returning to the base (Tazovskii) from the tundra. This situation was reported to us as being rather standard for the whole of the Russian north and most of Siberia.

For these reasons, the only reliable method of helicopter navigation (assuming that a GPS is not used) is by sighting surface landmarks and making use of an external or internal map featuring their relative positions – a practice that lies at the heart of visual piloting. The formal instructions[47] for this method of navigation are based on the assumption that an artificial (geographic) map is employed. According to these instructions, before takeoff the navigator should draw a course on an artificial map between the current position of the helicopter and the flight destination.[48] In the case of a short or middle-distance flight, the course is drawn simply by connecting the two points with a straight line.[49] The length of the flight in kilometers, its probable duration in minutes and its exact direction in degrees on a magnetic compass should then be calculated. The navigator then proceeds by determining the appropriate landmarks along the course. These can be curves of the river beds, streams, hills, islands of forest, individual trees, and other landmarks that are easily visible from above. It is recommended that the determined landmarks should be situated no more than 8–10 kilometers from one another in order to ensure that each successive landmark will appear in the visual field of the pilot before the previous one sinks behind the horizon. These landmarks are marked on the map and the distances between them in kilometers and flight minutes are specified. The prepared navigation plan (*shturmanskii plan)* should be confirmed by the first pilot-commander. Only after this can the helicopter take off. During the flight, the commander keeps the helicopter on course using a magnetic compass, while the navigator checks that the predetermined landmarks pass beneath the engine and suggests a change of course if necessary.[50]

---

47 The instructions referred to here come from a hand-typed, undated document titled *Instruktsii po poradku poletnoi navigatsii i poletnye skhemy po Tazovskomu OAO* ("Instructions on the order of flight navigation and flight schemes for the Tazovskii United Aviation Division"), which was kindly provided by one of our informants. Although the document has no date, the reference to Tazovskii OAO (here – *ob'edinennyi aviatsionnyi otrad* – "united aviation division") shows that it was produced before the privatisation of helicopters and probably before the collapse of the USSR. To the best of our knowledge, no newer formal instructions for navigation have been produced.

48 Maps with a scale of 1:200,000 (one centimeter = two kilometers) – called *dvuhkilommetrovka* – are normally used for this.

49 For longer flights, the change of magnetic inclination during the flight should be taken into account, which makes flying in a straight line impossible. Special "flying schemes" (*poletnaia skhema*) are used for drawing the course in this case.

50 It should be noted that keeping the helicopter on course at all times is difficult due to the influence of external factors, most notably crosswinds. Even a weak wind blowing constantly to one side of

Clearly, both determining the appropriate landmarks and correctly identifying them during the flight are absolutely crucial to visual helicopter piloting. Both of these tasks are, however, by no means trivial. Given the rather changeable climatic conditions and seasonality of the Arctic, it is always rather difficult to predict if a certain surface feature will be visible from the altitude of one kilometer. It is even more difficult to imagine what it would look like. For example, in winter, when the tundra is covered by snow, it appears from a helicopter like an absolutely plain and featureless surface which, to make the things worse, blinds the pilot with an incredible amount of reflected light. It indeed takes a lot of skill to recognise any landmark whatsoever on it.[51] However, even if the surface features are clearly visible, quite a lot of skill is still required to identify those that correspond to landmarks marked on the map. These skills, according to the older pilots, can be achieved only through long-term study of the landmarks and the memorisation of what they look like in different seasons and under different weather conditions. In other words, skill comes with experience that is possible to obtain only after many years of flying as navigator.

Before he obtains these skills, the navigator can easily miss or misjudge a landmark and, therefore, fail to fix the position of the helicopter correctly during the flight. His mistake usually becomes known fairly soon, when the next expected landmark fails to appear at the expected time. The same thing can also happen to an experienced navigator if, for example, the helicopter flies into a snowstorm or heavy fog, both of which can make the surface temporarily invisible. In such cases, the instructions advise the navigator to report immediately to the first pilot that they are off course. The commander is then advised by the manual to ascend "closer to the static ceiling" (1500 meters), hand over the operational functions to the second pilot, carefully observe the surface, and "try to apply his skills and experience to return the craft to the correct course." The commander is expected to do this by finding any familiar landmarks on the surface and recalling their spatial relation to the destination of the flight or to the landmarks along the lost course. Using this information, he is to infer how the helicopter can be returned to its proper course. In other words, because navigation with an artificial map performed by the second pilot has failed, the commander is expected to try to correct the error by using his mental map, which, the instruction appear to assume, he has developed from his long experience with visual navigation over the course of his career. The validity of this expectation was confirmed by the pilots interviewed: all of them reported that an experienced pilot, simply by looking

---

a helicopter eventually causes it to diverge from its initial course. This divergence usually cannot be determined, let alone evaluated by the helicopter equipment, rendering the observation of landmarks and course correction very important.

51 Visual navigation in winter – called "flying the white one" (*letat po beliaku)* by the helicopter pilots – was reported by the informants to be the most difficult task for a navigator. The ability to "fly the white one" without assistance from the commander demonstrates that the navigator has all the skills necessary for his job (but not for the job of commander, as will be explained later).

at the land surface, can tell immediately where the helicopter is and where it should fly in order to reach a certain destination. In their opinion, it was exactly the ability to do that within a large part of the Tazovskii region that, among other things, differentiated a commander from a good navigator. For the latter, the ability to recognise landmarks correctly was enough.

This evidence seems to show that the long experience with visual helicopter piloting that helicopter pilots used to acquire before the introduction of GPS technology allowed them to develop extremely large cognitive maps specifying the relative positions of a number of surface landmarks. These cognitive maps enable pilots, if necessary, to dispense with an artificial map as an instrument of navigation and to rely solely on their spatial cognitive capabilities. An interesting fact about these capabilities is, however, that they seems to be very specific in their application. Indeed, interviews with representatives of other professional groups in the Taz tundra showed a common belief that helicopter pilots have very poor navigation abilities when they either walk in the tundra or travel by any means of terrestrial transport. The following excerpt from an interview is typical:

> I would advise you not to go hunting with a pilot. Or, if you do, do not allow him to lead the way. The guys cannot orient themselves on the surface at all, but they always believe they can and, therefore, always lead you into a bog or deep river. They always think that certain places are closer than they actually are and forget about the bogs and streams along the way. They cannot recognise a place even if they say they know it. (Interview with I. Saitov, Tazovskii, April 2007)

Most pilots understandably dismiss this assumption and argue that with a little training and experience (which, however, many pilots admittedly lack) one can effectively apply helicopter navigation skills to finding one's way "on the ground." Nevertheless, they still tend to agree with the following statement, which is of great importance to our argument: the skills of visual helicopter navigation, even if they are so great that they allow a pilot to fly a helicopter without an artificial map, do not by themselves produce an equal or comparable navigation capacity when traveling on the ground. By way of explaining this phenomenon, many pilots observed that "things look rather different from above." This explanation can indeed be taken literally; it is rather probable that the pilots, who become accustomed to seeing particular landmarks from an altitude of hundreds of meters, simply fail to recognise them from the ground and, therefore, cannot fix their position on their cognitive maps. Even if this were indeed the only problem, it would still demonstrate an important fact about the spatial cognition of the pilots: their cognitive maps are not a product of their own perception only, but rather of a specific system in which a helicopter with its specific capacities constitutes an inseparable element. Furthermore, their spatial cognition can properly function, that is, be effectively used for solving cognitive tasks, only within this system. If it is to be applied to another system that does not contain a helicopter as an element, it has to be at least modified by learning what the elements of the cognitive map look like on the ground. However, a comparison of the navigation practices of pilots to those of reindeer herders shows that the matter may well be much more complicated than that.

4.3 Navigation practices of reindeer herders when traveling by reindeer sledge

Traditionally both Nenets and Komi reindeer herders performed the majority of their journeys outside their nomadic camps on reindeer sledges. A number of ethnographers (Golovnev 1995; Khomich 1995) have observed that Nenets reindeer herders do not like walking in the tundra, and often use reindeer sledges even for trips of less than one kilometer. This corresponds well to our observations among the reindeer herders of the Taz tundra. Although Komi reindeer herders do not seem to have an aversion to walking in the tundra (it is not uncommon for Komi to make walking trips of 10 to 15 km), they also definitely prefer to use a reindeer sledge in most cases. Nowadays reindeer sledges are increasingly being replaced by snowmobiles among both groups in the winter period. In the summer, however, reindeer sledges still represent the only means of transport available to the reindeer herders. As snowmobile navigation seems to have important particularities, it will be discussed in a separate section of this chapter.

Although the Taz Nenets can use marked tracks (for example, seasonal roads, riverbeds, or trails left by other reindeer sledges or snowmobiles in the snow) when traveling on a reindeer sledge, they do so only rarely. In most cases, a trip on a sledge involves blazing a completely new trail over the landscape. Only in a very few cases was it observed that reindeer herders used modern navigation equipment (GPS devices) during their trips on sledges, which is probably not surprising, as the use of GPS devices by private individuals without a special license was forbidden in Russia until December 2006. What is more interesting, however, is that the use of more traditional navigation equipment (map, compass) was never observed, despite the fact that some reindeer herders had these items in their tents. The administrators of the reindeer herding enterprise reported that their attempts to introduce this equipment among the herders had failed.[52] It can be said, therefore, that herders, a few

52 Two such attempts were reported. The first was made in 1982 when two reindeer herding groups (*brigades*) were moved to a new territory as part of a "rational pasture-use plan." The camps experienced substantial losses of reindeer and several dangerous accidents reportedly because of their lack of familiarity with the new territory. In order to help them, the administration provided the herders with detailed maps of the territory and attempted to give them a training course in orienteering with map and compass. However, the herders rarely used this equipment and gave it up completely after a few years, when they had become familiar with their new territory. The

cases of GPS use notwithstanding, still rely exclusively on their cognitive maps for navigation.

Both the interviews with and the participant observation among the Nenets reindeer herders indicate that inferences from a mental map play a much more important role in their navigation than following visually memorised routes. They described navigation in the tundra as traveling from one "known place" to another using routes that are not themselves memorised: in most cases, the herders insisted that they remembered neither the general view (vista) as it opened up from the sledge nor any of its details (such as hills or bushes) on the way from one "known place" to another. Furthermore, they insisted that it was impossible to check the direction of travel until the "known place" was reached. As one herder put it, "If you arrive at the place you wanted, then your course was right. Before that you never know." This assertion was supported by participant observation: in several cases, the herders lost their way (or believed they had) during the journey and, as conversations with them revealed, did not know (or at least were not sure) where they actually were until some "known place" was reached. However, the clearest evidence for the prevalence of mental-map-based navigation among the reindeer herders is their heavy reliance on methods of direction keeping, which they consider to be the most essential aspect of traveling by sledge. Among the Nenets, two such methods were observed in practice, although some herders reported the existence of other methods. The first method, used mainly in summer, spring, and autumn, relies on wind direction. When taking a course from a known place (evidently inferred from a mental map), a sledge driver takes note of the direction from which the wind is blowing on his body or, to use the herders' terminology, he "catches the wind." During the course of further travel, the herder always tries to guide the sledge in such a way as to keep this direction constant. This evidently enables him to keep to the taken course unless the direction of the wind changes – an event that, according to the reindeer herders, is the main reason for getting lost. The wayfinding directions that reindeer herders give one other can include a reference to this method, especially if the conversation takes place inside a tent or house:

> Vladimir: Where is Nikolay standing now? [i.e., Where is reindeer herding brigade no. 3, headed by Nikolay Hudi, camping at the moment?]
> Nemechi: On Hora-Sede. After tea, catch the wind in your right ear and carry on for three or four *nedalavas*.[53] You will be there by nightfall. Their herd is large, you cannot miss them. (Recorded in Taz tundra in July 2006. Informants: Vladimir Iar and Nemechi Tyseda)

---

second attempt was made at the beginning of the 1990s in an effort to cut down on the costs of helicopter transportation by preventing protracted searches for reindeer herding camps in the tundra. The herders were given maps of the Taz tundra and asked to report by radio the approximate geographical coordinates of their current position each time a helicopter was due to fly out to them. Despite some training again being provided, the misreporting of coordinates was so common that the idea was soon abandoned.

53 A *nedalava* is a Nenets unit of measurement that will be discussed later.

If the conversation takes place outside (which is most often the case), reindeer herders usually just point in the intended direction and tell the estimated length of the trip.

The second method of direction keeping, used in winter, is based on *zastrugi* – sharp grooves or ridges carved by the wind on the surface of the snow. The orientation of *zastrugi* is, therefore, parallel to the prevailing winds. In the Taz tundra they most often stretch roughly in a north-south direction (with some inclination to the east), deviating from this general pattern near hills or large objects standing out from the surface. At the start of a trip, a reindeer herder touches the snow surface passing under his sledge with his feet. This allows him to feel how the *zastrugi* hit his feet as the sledge crosses them, and, therefore, to feel the angle at which the *zastrugi* are crossed. During the trip, the herder checks to make sure this angle has not changed by extending his leg out from the moving sledge and touching the surface from time to time.

A practice similar to "catching the wind" was also often mentioned by Komi reindeer herders and said to be an important part of their navigation routine. Their use of this practice was also observed in a number of instances in the course of fieldwork. Another method of course keeping mentioned by Komi reindeer herders involved observing the position of the sun. As one herders explained, "Note how the sun is situated toward you, and go in such a way as to keep its position constant." This method was never mentioned by the Nenets reindeer herders. On the other hand, the use of *zastrugi* for the purpose of course keeping was neither mentioned by nor observed among the Komi.

Needless to say, all the methods of course keeping described here would be redundant if a traveler were to follow a visually memorised route in the stimulus/response manner. Thus, the existence of these methods and the great importance ascribed to them by our informants suggest that an essential part of their navigation practice consists in following a course inferred from a mental map rather than a visually memorised route and updating this course through a new mental map lookup every time one reaches a "known place". This is hardly surprising given that the reliance on mental maps obviously has important advantages to the tundra dwellers. Most notably, it allows them to navigate the tundra in conditions of poor visibility, for example in darkness (which is especially important during the polar nights) or in fog, which is a rather frequent phenomenon in summer and autumn. Moreover, it ensures that the traveler is able to take the shortest and the most straightforward route to the destination. This is obviously important because travel conditions in the tundra tend to be rather harsh during the most of the year.

In a sense, the navigation practice described above closely resembles what Ranxiao Wang (2012) has recently called the "egocentric-updating-and-reload model with enduring allocentric representation." Indeed, in the "known places," the allocentric[54]

54 The term *allocentric* signifies that the reference points are independent of the position or perspective of the traveler.

representation of the environment (i.e., the mental map) gets aligned with the egocentric reference frame, which is defined by the angle between the traveler's heading and the direction of the wind or of the *zastrugas*. During his or her travel between "known places," the traveler relies mainly or solely on this egocentric representation, which gets constantly updated by monitoring the changes in the angle between the heading and the wind direction. Finally, once a new "known place" is reached, the egocentric representation gets "reloaded" by re-aligning it with the mental map. In this model, the mental map lookup essentially consists of the alignment between the egocentric and allocentric frames of reference by means of a position fixing (representing the position of the traveler in relation to other locations on the mental map, including the target of travel) and inferring the egocentric bearings toward these locations and the target (Wang 2012: 579; see also Easton and Sholl 1995). Both of these operations can be easily performed if the locations on the mental map are represented with reference to some allocentric axes or bearings that, furthermore, can be somehow traced in or imposed upon the world as it is actually perceived by the traveler (cf. the 'absolute reference frame' as discussed in Levinson 2003). A good example of such allocentric axes is the cardinal bearings (north, south, east, and west). If the relative positions of locations on the mental map of a traveler are encoded with reference to such bearings, then starting his or her trip from a "known place" or having reached a "known place" in the course of the trip, the traveler can fix his or her position vis-à-vis the target as, for example, "the target is to the northeast of me." Once his or her position is fixed in this way, the traveler can easily take the course toward the target, given that he or she has the means of determining where north, south, east, and west are in the world he or she actually perceives. This can be achieved, for example, by paying attention to the sun, the stars, or the arrow of a magnetic compass.[55]

A fair amount of evidence from our fieldwork suggests that Komi reindeer herders rely on exactly these methods of position fixing and course taking. Indeed, cardinal bearings[56] were often invoked when the reindeer herders described to us the relative positions of different locations in the tundra. More importantly, we observed that these bearings were frequently used in conversations the herders had with one another. In our opinion, this suggests that Komi reindeer herders indeed represent the relative positions of locations in the tundra with reference to the cardinal bearings.

---

55 Of course, the cardinal bearings north, south, east, and west are not the only ones that can be used for the purpose. For example, inhabitants of the island of Bali represent locations with reference to bearings called *kangin*, *kauh*, *kaja*, and *kelod*. While *kangin* and *kauh* more or less correspond to east and west, *kaja* signifies the direction toward the mountains (away from the sea), while *kelod* is the direction toward the sea (Wassmann and Dasen 1998; see Levinson 2003 for further examples and discussion).

56 In the Komi language, cardinal bearings reflect the course of the sun during a day and, therefore, coincide with those in European languages. Thus, the four main bearings are *lunvyv* (lit. "midday," i.e., south), *rytvyv* ("evening," i.e., west), *vojvyv* ("midnight," i.e., north) and *asyvvyv* ("morning," i.e., east).

We also found that male Komi reindeer herders try to constantly keep track of the cardinal bearings and, furthermore, consider the ability to do this an important sign of maturity. The informants insisted that a grown-up male reindeer herder should always be able "to point to the north," and that this ability, among other things, differentiated him from a woman or a child. We observed that male Komi informants were indeed able to do this most of the time, inside the camp as well as while traveling. When adult reindeer herders were accompanied by children in their travels, they frequently asked boys to point to the north. If the boys could do this, they were praised by acknowledging their progress toward maturity with comments such as, "You will be a grown-up soon," or "You will need to look for a bride soon." The concrete methods the herders employed to keep track of the cardinal bearings could not be determined: most of the herders denied that they used sun, stars, or any other environmental indicators to do this, and insisted that they simply "knew" where north was. This knowledge, they claimed, was acquired through experience. Regardless of the cognitive mechanism underlying this ability, the attention male reindeer herders pay to it suggests that it plays an important role in their everyday lives.

Among the Nenets herders, the methods of position fixing and course taking are likely to be different. Although it has been claimed by several researchers (e.g., Golovnev 1995; Khomich 1995; Spodina 2001) that the Nenets have developed a system of cardinal bearings based on the course of the sun,[57] we found no evidence of this system among the Taz Nenets reindeer herders. Indeed, fourteen out of sixteen Nenets informants whom we asked to name the cardinal bearings in Nenets could not provide terms for any bearings other than north (*ngerm* in Nenets). That term, however, was part of the name of a regional newspaper (*Naryana Ngerm* – lit. "Red North") and of the local folklore ensemble (*Ngerm haiar* – lit. "Northern Summer"). Many of our informants admitted that they knew this term only because it appears in these names. Most informants knew Russian terms for cardinal bearings, which they learned during their compulsory education in the boarding school in Tazovskii. Still, only four informants were able to point to these bearings with an error of less than 45 degrees.[58] Most informants claimed that they did not know these bearings in the "real world" and, therefore, could not point to them. Several in-depth qualitative interviews with reindeer herders were performed in order to try to detect the existence of any other system of allocentric bearings or axes that they might use. No evidence of the existence of such a system was found. Although it is possible that some system of this sort still exists and our failure to find it is related to the objective difficulty of inquiring

57 According to these researchers, this system includes the concepts of the four basic bearings (north, south, east, and west), as well as a number of more specific bearings, such as NW and even NNW.

58 In the context of the previous discussion of position fixing and course taking among the Komi, it is interesting that one of the informants who was able to point to the cardinal directions relatively accurately was a young woman.

into abstract phenomena of this kind, it is more likely, in our opinion, that the Nenets reindeer herders do not represent the relative position of locations with reference to any sort of allocentric bearings.[59]

Although the principles of representation of the relative position of locations on the Nenets reindeer herders' mental maps require further investigation, our interviews suggest that they represent both directions and distances from each location to other locations. The former is achieved, most probably, by memorising those segments of a vista, as it opens out from a particular location, that correspond to the directions toward other locations, as the following statements clearly illustrate:

> Sem'on: Yes, it is true. I do not know where north, south, and so forth are. Why do I need to know this?
> K.I.: How then do you remember in what direction you should go to reach some place?
> Sem'on: Oh, it's easy. I just remember the directions from here to other places.
> K.I.: How?
> Sem'on: I remember how I should start. Say I want to go to Puhutsa-To [the name of a large lake – *K. I.*]. I know that in order to do that I should start toward that hill [points] and pass across its slope a little bit to the left of its top. In order to reach Sangibei [the place name], I should start toward that big bush on that slope [points]. I start in that direction, catch wind as I go toward the bush and then continue by wind. And I am there. Very simple, is it not? I do not need all these norths, souths, compasses, whatever.
> (Interview with Sem'on Habdu, recorded in Taz tundra in June 2009)

Similar descriptions have been provided by other herders. They suggest both the principle of the organisation of their mental maps and the mechanisms of position fixing and course taking they use. Essentially, the mental maps of Nenets seem to be interobject relationship models that take the form of multiple "egocentric snapshot spatial representations" (see Wang 2012: 578), each of them related to a particular "known place". It can be noted that this form of mental map has a set of serious limitations. Thus, it is very difficult to add a new location to a mental map of this form, that

59 One piece of evidence supporting this conclusion is the way Nenets reindeer herders deal with artificial maps. When we asked our Nenets informants to show us the positions of certain locations on a map, the informants would usually rotate the map in their hands or put it on the ground and walk around it, obviously checking how the scatter of landscape elements on the map looks from different points of view. Some informants commented that they should first understand how the "land" was "depicted on the map." When the informants proceeded to show us the locations we had asked for, they would often hold the map in their hands sideways or upside-down rather than in its "normal" orientation (with north on top). No regularities could be observed with respect to the position in which the map ended up being held. Furthermore, in a number of cases the informants indicated some of the locations while holding the map in one position, but then rotated it in order to find and indicate other locations. Although this behaviour can be interpreted in several ways, it most likely suggests that aligning the scatter of elements on the map with any stable allocentric axes did not help the informants recognise the locations, that is, it did not help them find the correspondence between the locations depicted on the artificial map and those encoded on their mental maps.

is, to learn the relative position of a new place. Indeed, in order to fully incorporate a new location into his mental map, a Nenets herder should, it seems, learn the bearings toward this location from all other locations already present on his mental map by discovering and memorising the relevant segments of vistas that open out from these locations. Secondly, the task of taking a shortcut or detour along a novel and unfamiliar route – arguably the main advantage of having a mental map as far as spatial navigation is concerned – becomes a rather non-trivial cognitive challenge if the described principle of mental map organisation is used.

Indeed, suppose that a Nenets reindeer herder is traveling from location A toward location B and that he wants to detour to location C from a certain point α on his track. While locations A, B, and C are incorporated into his mental map, point α is not: the route from this point to location C is completely novel for the herder and has never been experienced by him. Given that the herder does not relate the locations to any general system of allocentric axes, the only way he can infer and perform this novel route seems to be by an integration of route legs not dissimilar to the famous dead reckoning of mariners. The herder has to integrate mentally the leg from α to either location A or location B with the leg from this location (A or B) to location C. From the resulting figure, he has to triangulate the line connecting α and C and evaluate the angle between this line and the line AB that he is currently on. Finally, he has to correct the course of his movement by this angle. Although all these operations are mathematically possible, performing them "in one's head" would demand quite impressive abilities of calculation and geometric imagination. This is particularly true if the legs to be integrated and the detour to be performed are long: in this case the evaluation of the angle for taking the detour course is, on the one hand, particularly difficult and, on the other hand, requires particular accuracy. Indeed, in the case of a long detour, even a small mistake in the initial course taking can result in "missing" the target location by a long distance.

These limitations suggest that there is something more to the representation of the relative position of locations and the procedure of position fixing and course taking than the Nenets informants were able or willing to disclose during interviews. On the other hand, if the herders indeed face these limitations, it might explain the peculiar regionalised structure of their mental maps described in the previous chapter. Recall that in the course of pointing experiments many Nenets herders could point to places within the region (*ya*) where they were currently situated, but not to places outside that region. After migrating to a new region, many herders became able to point to those locations within the new region that they were previously unable to point to. On the other hand, they became unable to point to locations in the region that they had left. If the representation of relative positions of locations among the Nenets herders is indeed based on representing bearings and distances between them, then these results can be easily explained by the difficulty of incorporating locations into their mental maps as described above. Indeed, it can be suggested that instead of memorising the bearings toward any location from all other locations they know (which is both virtually impossible and unnecessary), the herders memorise only the bearings to and from locations that are situated close to each other and make up a cluster. These clusters are then conceptualised as regions (*ya*). The bearings from the particular locations inside a cluster to those outside it are not memorised; rather, the herders can memorise bearings toward the other regions as wholes (for example, by recalling the bearings toward locations situated on the borders of these regions but still within the region they are currently in). Given that the regions are not particularly large (their average size, as noted in chapter 3, is approximately 20 x 20 km) and, therefore, that they do not include a very large number of locations, this structure of mental maps allows one to solve the problems of incorporating new locations and having to keep in mind an enormous number of bearings from any location to all other locations (*cf.* the discussion on the relational mental map's simplification in Wang 2012: 579). Furthermore, it can also be used to keep the complicated procedure of inferring and taking a novel route within manageable limits. Indeed, in the case of taking a novel route inside a particular region, a Nenets herder is unlikely to face the complicated problem of integrating long legs; furthermore, he does not have to determine the angle of the course to be taken with precise accuracy because the detour is likely to be short.[60] However, in the case of traveling to a place situated in another region, whether as part of a detour or otherwise, the herder would have to take the general course toward this region – probably to some known place on its border – and, once the region is reached, correct his course through a new position fixing.

As the discussion above clearly suggests, an important element of navigation when traveling by reindeer sledge is evaluating the distance that has already been covered and the distance that still needs to be covered. In the case of inferring and performing a novel route, this evaluation is important for representing one's position in relation to the locations on the mental map. For the Nenets, as explained above, this allows the representation of legs to be integrated. The evaluation of distances covered is also indispensible for the herders to understand when they are about to get lost. It is also likely to play an important role in place recognition, as we shall demonstrate shortly.

Both Komi and Nenets reindeer herders evaluate distances using specific units of measurement – called *nedalava* in Nenets and *duk* in Komi – that correspond to the distance the reindeer can pull the sledge without stopping for a rest. Nenets usually say that a *nedalava* is roughly equal to ten kilometers, sometimes adding that it is "more like eight kilometers in the summer and twelve kilometers in the winter when

60 In the relatively flat Taz tundra, one can normally see objects and landmarks from a distance of 3 to 5 km (up to 7 km on a clear day), as these objects or landmarks stand out clearly from the monotonous environment. Thus even if a herder were to miss his target by this distance due to a mistake in course taking, he would still be able to see it at the end of the journey. Of course it should not be forgotten that the conditions of visibility deteriorate significantly when there is heavy fog or a snowstorm, and/or during the polar night.

it is easier for reindeer to pull the sledges because of the snow." This remark shows that the actual length of the unit depends on a number of factors which, apart from the season, include landscape, surface conditions, and the condition of the transport reindeer. Interviews and observations clearly indicate, however, that sledge drivers memorise the distances between "known places" in these units.[61] This gives them the possibility of determining the approximate distance still remaining to reach a destination by the number of stops already made. For example, if the distance between point A and point B is known to be four *nedalavas* or *duks*, then after the third stop a reindeer herder knows that he should be already quite close to his target. If the target does not appear after the fourth, fifth, and sixth stops, then he can be pretty sure that something has gone wrong and that he has gotten lost.

What is more important for our discussion, however, is that distance knowledge plays an essential role in guiding the herders' expectations and attention which, as psychologists well know, are integral parts of the recognition process. Thus, to use the example above, if a place looking very similar to B is encountered after two stops on the way from point A, a traveler would probably decide that this place, whatever it looked like, could not be B, because it was too soon for B to have been reached. On the other hand, if four stops on the way from A have already been made, the traveler may well decide that an encountered place is in fact B even if it does not look exactly as he remembered it. The evaluation of the distance traveled thus provides a clue to place recognition, and this clue seems to be rather important for the herders: it was observed that some Nenets informants had problems recognising places depicted in photos even when those photos were taken during a journey with the same people and at the same spots from which the herders had successfully recognised the places during their actual travels. Further evidence and the implications of the importance of this clue will be discussed in the next section. For the present, it is enough to state that from the viewpoint of navigation analysis, the information that "point B is approximately four *nedalavas* from point A" represents, in our opinion, the essential characteristics of points A and B. These characteristics are just as important as the visual characteristics of these places.

In order to complete our account of navigation on a reindeer sledge, it should be mentioned that, according to both Nenets and Komi reindeer herders, one of the most important qualities of a successful navigator is that of bravery. The informants commonly insisted that a coward always gets lost in the tundra. The following quotation from a Nenets reindeer herder provides a good explanation why:

> The main thing you need [in order not to get lost – *K.I., M.J.D.*] is to be brave and to go forth. It is not easy. When you travel in the tundra, you always think, "Have I taken the right direction?" and, "Have I not missed the place where I am going?" Everyone has these fears, especially if you believe that you should have already reached a place but

61 This makes sense when you consider that a sledge traveler does not have the technical means of measuring distances in absolute units.

> you cannot see any sign of it around; then, these fears become really strong. But you should not surrender to these fears. You should be brave! It is not easy, especially when you are alone in the darkness. You can think, for example, "I have probably gone too far to the left; I should go a little bit to the right of the course I am taking now." You can even eventually become completely convinced of this, especially if you do not see the place when you think you should already be there. Still, you should not change course. If you keep on the same course, you will eventually reach someplace, maybe not the place you wanted, but still a place you know. However, if you indeed change course at least once, you will get lost, because a little bit later you will start to believe that the new course is also wrong and you will change it again, then again. If you start changing course you will be unable to stop, believe me, nobody can. Then you will start to go in circles until your reindeer fall, and then you will walk in circles. All the people who have gotten lost and died in the tundra did so because they were not brave enough and surrendered to their fears. (Interview with Nemechi Tyseda, Taz Tundra, June 2006)

The lack of bravery and confidence is believed by reindeer herders to be the main reason why some people cannot orient themselves in the tundra. A certain number of such people are also present among reindeer herders. The following quotation from another Nenets herder concerns one of those people:

> (Name of a reindeer herder) always gets lost wherever he goes. Do you know why? Once he got lost and wandered in the tundra for several days before we found him. Maybe he was drunk or something like that. After that he became a coward. He goes somewhere and once the tent disappears from sight he thinks, "I have gotten lost again! Of course I am going in the wrong direction! I should go a little bit to the right." And so he does and he can do nothing about that. Then after some time he thinks, "I have taken a wrong turn again! I am lost!" And he goes in a new direction once again. And of course he gets lost – he goes in circles! And he can do nothing about it. Nobody can. (Interview with Iela Salinder, Taz tundra, July 2006)

This description suggests that the practice of navigation on a reindeer sledge, similar to visual helicopter piloting among experienced pilots, is based on a mental map from which direction and estimated travel distance are inferred. During the journey, direction is maintained by certain techniques (including "catching the wind" and, among Nenets, checking the *zastrugas*). However, we would argue that, as in the case of visual helicopter piloting, understanding these maps and techniques alone is not enough to completely account for spatial navigation. The description of yet another wayfinding practice will help demonstrate this point.

4.4 Navigation during a snowmobile journey

At the moment, snowmobiles are probably the most widely used means of transport in both the Bol'shezemel'skaya and the Taz tundras during the snow period. Widespread use of snowmobiles started in both regions toward the end of the 1980s, although the

Nowadays Burans can occasionally be used by reindeer herders also during the snowless period along with reindeer sledges. This is, however, quite damaging for the snowmobiles. Photo: Kirill Istomin.

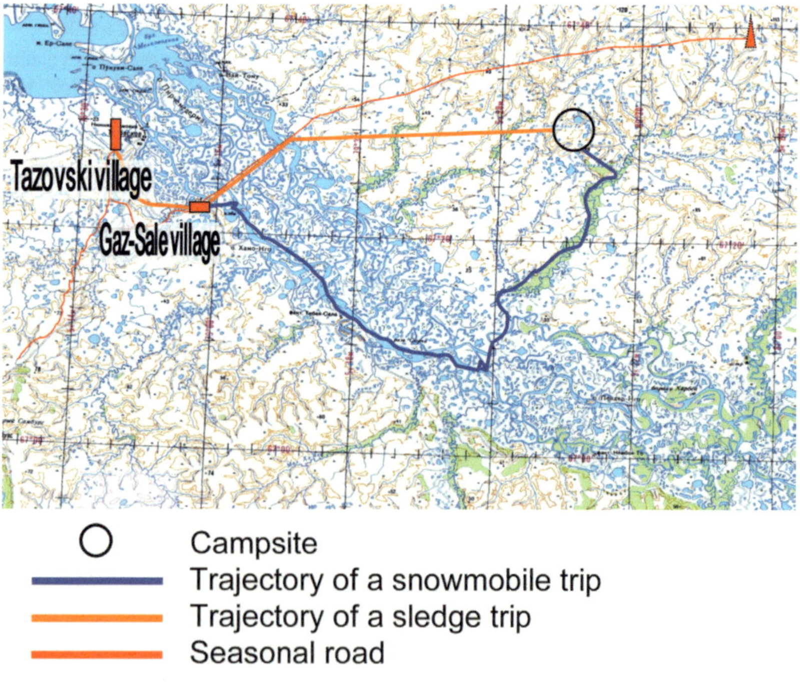

Fig. 4.1 Examples of reindeer herders' sledge and snowmobile journeys between their campsite and the settlement of Tazovski (based on GPS recordings).

occasional snowmobile had been in use since the beginning of that decade. Nowadays, snowmobiles are used by the settled population, and are a popular alternative to the reindeer sledge among nomadic reindeer herders and (in the Taz tundra) semi-nomadic fishermen. Most snowmobiles used in both regions are the famous Russian-produced Buran brand, first built at the end of the 1970s. The construction of these snowmobiles is now rather outdated despite several attempts to modify it throughout the 1990s, and the vehicles are rather unreliable by modern standards and prone to break down at any moment. Nevertheless, most snowmobile users prefer Burans to imported snowmobiles because of their low price, availability of spare parts and, most importantly, the users' familiarity with the Burans' construction and engine, which allows most drivers to repair their Burans themselves in the tundra. This is something they cannot do with imported snowmobiles, which are more reliable but nevertheless frequently break down when one is alone in the tundra.

Participant observation in the Taz tundra showed that snowmobile trips, when compared to trips by reindeer sledge, are much more often carried out along marked trails: in fact, the majority of the Buran trips observed or reported were made at least partly along frozen rivers, seasonal roads, or tracks left by other snowmobiles or reindeer sledges. More often than not, following these trails resulted in deviations, sometimes rather significant, from the shortest way between the point of origin and the destination. This holds true for snowmobile trips undertaken by representatives of all groups (settled, semi-settled, and nomadic), although nomadic reindeer herders undertook more trips that did not follow marked tracks and were of greater distance than representatives of either of the other two groups, and semi-settled fishermen, for their part, took more such trips than did representatives of the settled population. It was also observed that some snowmobile drivers, particularly those from the settled groups but also several reindeer herders and semi-settled fishermen, did have navigation equipment (compass, map, and/or GPS device) with them during their journeys. However, the use of this equipment for navigation purposes was observed only among representatives of the settled groups.

As snowmobile journeys are often taken along established tracks, it is highly likely that snowmobile drivers rely at least partially on route knowledge. This reliance was indeed evident among many representatives of the settled groups, as well as among the semi-nomadic fishermen, who, as was mentioned in the previous chapter, could point neither to the destination of their planned journey nor to many (or sometimes, any) points along their way, despite being able to describe the route.[62] This suggests that these points were not integrated into their cognitive maps by means of a mental map, and the only way they could navigate, given that navigation equipment was not

62 For example, one fisherman described his route as follows: "We will follow the Vane-porod [one of the river beds forming the Taz river basin] to the mouth of the Puhuts-porod. We will then cross over to the Puhutsa-to." This informant, however, was unable to point to any of the mentioned places, arguing that the route to them is curved.

used, was by means of route knowledge. This also suggests that these drivers could not actually evaluate the degree to which the routes of their journeys deviated from the shortest possible way between the origin and destination. The majority of reindeer herders, on the other hand, were able to point accurately to their destination and/or to some of the points on the shortest route to it, suggesting that theoretically they could navigate this route using a mental map. This did not, however, stop many of them from using marked tracks during their actual travel, thereby causing significant deviation from this shortest possible route. Informants (reindeer herders and semi-settled fishermen) provided a number of objective explanations for this behaviour. First, a snowmobile is much less capable of blazing a new trail through deep snow and over difficult terrain than a reindeer sledge. The iced-over rivers and already well-trod tracks allow faster travel and decrease the danger of getting stuck during the journey. Furthermore, since the snowmobile is faster than a sledge and does not become tired as a reindeer does, one can deviate from the shortest route to follow a better track and still accomplish the journey in reasonable time. Finally, a journey by snowmobile involves careful planning of fuel supply and use. Using flat ice and trodden tracks is more fuel efficient. Moreover, the places where one can refuel tend to be concentrated along rivers and seasonal roads. Nevertheless, in some of the cases we observed, the deviation from the shortest route caused by using marked tracks during a snowmobile journey was so significant that it offered no economy in terms of time or fuel expenditure, and the places where a driver could have refueled were not visited (Figure 4.1). This gives good grounds to seriously consider another group of reasons for following rivers and established tracks when traveling by snowmobile. A number of the older reindeer herders reported that navigating the tundra on snowmobiles is much more difficult for them than doing so by reindeer sledge:

> In the first place, on a Buran you cannot keep your course by using wind direction. It goes so fast that it is difficult to "catch the wind" – it feels like the whole force of the wind is always blowing in your face, right? So you cannot check whether or not you are going in the right direction. Also, on a Buran you can get to a known place much more quickly, but by the same token you can miss the place much more quickly as well. You can confuse places much more easily.

When asked why this is so, the herders would answer, "Because the world looks different from a snowmobile than it does from a sledge."

The first point that this quotation highlights is a problem that exists when one tries to apply the practice of sledge navigation to traveling by snowmobile. Indeed, the speed of travel creates such a strong flow of air in the face of a driver that catching the "natural" wind without stopping becomes impossible. One can easily anticipate that the method of keeping direction by feeling *zastrugas* would also fail unless the angle between the direction of travel and the *zastrugas* is evaluated visually (which is impossible most of the time in winter due to poor visibility). Navigation with a snowmobile therefore demands new methods of course keeping that are different from those used by sledge drivers. One such method was reported to us in 2009: starting from a known place, a driver chooses a star resting low on the horizon in the direction in which he wants to set his course and steers directly toward that star. Clearly, this method can be effective only when the distance between "known places" – that is, places where the position can be fixed and the direction of the travel can be updated – can be covered in less than about two hours: in this case the slight change in the position of the star during the trip will not result in an error that will send the driver many kilometers off his course.[63] This explains why this method is not used when traveling by sledge in the Taz tundra.[64] However, in the case of high-speed transport such as a snowmobile, the method can be practical. Still, if the distance between known places is significant, relying exclusively on the stars can lead to a significant deviation from the route. This indeed justifies inferring only shorter sections from the mental map and relying more on route knowledge.

The second point highlighted in the quotation is, however, much more difficult to understand. It suggests that using a snowmobile has a negative impact on the correct recognition of a place. As noted in the discussion of visual piloting, helicopter pilots have the same problem when they try to apply their cognitive maps to traveling on the ground (interestingly, they describe the phenomenon in essentially similar terms: "Things look different from a helicopter"). In the case of reindeer herders, however, it is hardly possible to explain this by reference to the different angles and perspectives. Indeed, although a snowmobile is a little higher than a sledge, this by itself cannot lead to a very different perception of a landscape. However, the point can be understood quite literally if one takes seriously the assertion we made in the previous section, that perceived distances between places represent an essential characteristic of those places that is no less important than their physical appearance. Indeed, as described in the previous section, Nenets sledge drivers memorise the distances between "known places" in *nedalava*s (i.e., the stops one has to make when moving between places in order to let the reindeer have a rest). In the case of a snowmobile journey, the distances memorised in this way simply do not make any sense. Therefore, they cease to

63 In Arctic regions the majority of the stars do not set, but move clockwise around the zenith, the rate of displacement being 15 degrees per hour. Therefore, after two hours, the deviation of the star course from the "real" course will reach 30 degrees.

64 Golovnev reports that Nenets reindeer herders of Yamal peninsula use this method during their sledge trips with a certain correction for star displacement: after completing each *nedalava*, they "shift sight to the next star to the left of the previously used star, and so on" (Golovnev 2009: 43). In our opinion, however, even with this correction the method cannot be very reliable, especially in the case of a long trip: the time it takes to complete a *nedalava*, as well as the angle between the previous star and the next one to the left, can vary too much. For these reasons, in the case of a sledge trip, this method may be used in addition to "catching the wind" (also mentioned by Golovnev) and feeling the *zastrugas*, but it could hardly replace them. In any case, our informants from the Taz tundra reported that they use this method only when they travel by snowmobile.

guide the herders' expectations and attention, which are crucial for correct recognition. To elaborate on the example from the previous section, if a place that looks like point B appears in sight after thirty minutes of snowmobile travel from point A, a reindeer herder who knows that the distance between these places is four *nedalavas* cannot determine if this place was encountered too early (and therefore cannot be point B), too late (in which case it is also not point B), or at the right time. Furthermore, he might miss the real point B, especially if he does not remember what it looks like in detail, because he does not know when it is time for it to appear and therefore does not pay attention. The informant quoted above quite astutely called attention to these effects when he said that one can both confuse places and miss them much faster despite (we would argue, because of) being able to get to a known place much more quickly with a Buran. When operating a Buran, a reindeer herder accustomed to evaluating distances in *nedalavas* is deprived of one of his most important clues for recognising places and, indeed, these places start looking very different to him. It can be supposed that Komi reindeer herders, who memorise distances in *duks* (analogous to the *nedalavas* of the Nenets), would experience similar problems. Unfortunately, no research to test this prediction has thus far been performed.

4.5 Discussion: Culture, spatial cognition, and navigation practices

The empirical data presented in the previous section show both the similarities and differences in spatial cognition and navigation practices between the two groups of reindeer herders, as well as between these groups and the helicopter pilots engaged in visual piloting. The most important similarity in all three cases is that navigation seems to be based mainly on inferences from mental maps (in the case of reindeer herders and experienced helicopter commanders) or artificial maps (in the case of helicopter navigators). In order to make these inferences and use them for wayfinding, the representatives of all three groups have to solve the tasks of position fixing, course taking, and course keeping. As far as these tasks go, their navigation practices seem to be based on a universal cognitive process.

This universality collapses, however, when one examines the details of the navigation practices and the cognitive structures (such as mental maps) that support them. Here, pronounced differences seem to exist among the three groups. Quite a number of these differences, such as the differences between the pilots and the herders in the size of the territory covered by their mental maps and the number of localities encoded in them, are differences in content and can be explained by the fact that the representatives of the groups experience territories of different sizes and store different information regarding them. Other differences, however, undoubtedly represent differences in cognitive processes themselves. This particularly concerns the procedures used for position fixing and course taking. In the case of the helicopter pilots, these procedures seem to employ cardinal axes and technical equipment (magnetic compass), while the Komi reindeer herders appear to rely on cardinal axes and knowledge about their position in the empirically perceived world, and, as far as we can judge now, the Nenets herders use no cardinal axes, but rely on their memory of particular sections of vistas corresponding to bearings toward certain locations. The procedures of course keeping seem to be more or less similar in the case of the two groups of reindeer herders as far as traveling on a reindeer sledge is concerned. In the case of the visual helicopter piloting, course is again kept by an entirely different method: through tracing and observing pre-determined landmarks.

The differences in position fixing and course taking between the three groups seem to be closely related to differences in spatial representations (mental maps). Again, these differences seem to go deeper than just the content of these representations and actually touch their very form. Thus, the procedures of position fixing and course taking employed by Komi reindeer herders suggest that they represent relative positions of locations with reference to cardinal axes. The same can, of course, be said about the artificial maps used by helicopter navigators as well as, most probably, the mental maps of the helicopter commanders. In contrast, the spatial representations of the Nenets herders, it seems, do not employ such axes. Rather, their mental maps appear to be divided into small regions. The bearings from each encoded place to other places inside the region are represented as sections of vistas that open out from this place. Finally, the spatial representations of the reindeer herders seem to differ from those of helicopter pilots with respect to the representation of distances. The special way the herders represent distances by memorising the number of stops one has to make on his way between locations seems to have important implications for the process of place recognition and contributes to the specific problems some of them experience when traveling by snowmobile. In all these instances, culture seems to "play variations" on cognitive structures and processes themselves rather than just on the content supported by these structures and processes.

How is this variation to be accounted for theoretically? We suggest that the best way to account for most of this variation is by invoking the theoretical approach of distributed cognition as proposed by Edwin Hutchins (Hutchins 1995) to explain the modern practices of marine navigators. In essence, the distributed cognition approach understands cognitive processes as being fostered in systems (called "cognitive systems") that are larger than those of an individual.[65] In other words, the cognitive process is performed and cognitive tasks are solved not by an individual alone, but rather by a system of elements, one of which is the individual. The other elements can be material objects (for example, special equipment for computing or observing), as well as other individuals (as in the case of teamwork during marine navigation). An individual's cognitive processes and representations are, in this case, only a part of the

65 See Clark 2010 for a recent extended discussion. See, however, Rupert 2009 for a criticism of this approach.

general process taking place within such systems, and can be understood only with reference to the system as a whole.

Employing this approach for analysing the empirical material presented in this chapter, we can conclude that a great deal of the navigation practices can be understood if we view them as performed in different cognitive systems. Each of these systems includes, in addition to the human individuals with their spatial representations, other elements, and both the number and properties of these elements differ significantly from one system to another. Thus, visual helicopter piloting is performed in a cognitive system that, in addition to the two pilots, includes the helicopter (which flies at a certain speed and altitude, produces certain effects on perception, makes it impossible either to feel the wind or to see or feel *zastruga*s), a magnetic compass, and an artificial geographic map. In turn, navigation on a reindeer sledge is performed in a system that includes a sledge, a reindeer team (with a certain regime of "run and rest"), wind, *zastrugas*, and probably a number of other elements. The cognitive system employed in snowmobile wayfinding minimally includes a snowmobile (again having a certain height and being capable of traveling at a speed that precludes the possibility of feeling the wind and *zastrugas*), and marked tracks (rivers, paths, etc.). All these elements are systematically related to one another during the wayfinding task. Furthermore, the ability to solve this task with reasonable efficiency can best be viewed as the emergent property of these systems as a whole, which is not possessed by any of their elements (including humans) alone.

Returning to the human component of these systems – that is, to the spatial representations of their users – we can conclude that these representations are 1) the product of the specific cognitive systems, and 2) designed to work in exactly these systems. Indeed, the primary reason for the frequent absence of bogs on the mental maps of helicopter pilots is, of course, that these maps were mapped from a helicopter, that is, in the framework of a system in which the bogs are both barely visible and unimportant for performing the cognitive tasks at hand. The presence of the magnetic compass and heavy reliance on it during the flight makes it both easy and important to relate the locations with reference to the cardinal directions. On the other hand, it is precisely the absence of bogs on the cognitive maps of the helicopter drivers (among other factors) that makes it difficult for them to move effectively through the tundra on foot. Similarly, no one but a person traveling on a reindeer sledge can represent the distances between places in *nedalava*s: a snowmobile driver, for example, physically cannot obtain such a representation. Yet it is precisely the representation of distances in *nedalava*s that creates problems for herders using snowmobiles.

It is probably worth noting that the statement above does not mean that spatial cognition is rigid and completely determined by the professional group to which a person belongs. In other words, we would not argue that a helicopter pilot cannot learn to navigate in the tundra on skis simply because he is a helicopter pilot, or that a reindeer herder cannot learn how to navigate in a helicopter. We would, however, argue that this would demand a certain adjustment of their cognitive maps to the new cognitive system, in other words, a re-mapping of their cognitive maps. In order to find his way in the tundra on skis, a helicopter pilot cannot simply apply the knowledge of a territory that he has acquired from flying a helicopter. Some changes to this knowledge in the form of acquiring new representations are necessary.[66]

We would like to note, furthermore, that culture plays an important role in determining both the elements and the structure of the cognitive systems in which individuals experience and navigate space. Indeed, it would be a truism to say that the reindeer sledge as a means of transportation is part and parcel of a reindeer herder's culture and that it is intimately related to other aspects of that culture and economy. It would also be a truism to say that a helicopter is both a product and a part of a settled industrialised culture. An obvious conclusion to be drawn from this is that one important way culture can "play variations" on cognition is through determining the parameters of cognitive systems in which certain cognitive tasks are routinely experienced and solved. This facilitates the solving of cognitive tasks within a particular cognitive system, and can hinder it within other systems. The examples of spatial cognition discussed in this chapter show that the parameters of different cognitive systems can lead to significant differences in the mental maps acquired and used. This, in turn, suggests that culture can have a profound impact on cognition.

Of course, not all differences in spatial cognition between the three groups can be explained by differences in cognitive systems. The differences between the two reindeer herding groups, particularly those concerning how the relative positions of locations are represented on their mental maps, cannot, it would seem, be accounted for in this way. We are forced to admit that we do not have a satisfactory explanation for these differences at this moment. Nevertheless, it can be safely concluded that a particular cognitive system, which is necessarily a cultural phenomenon, is likely to have a pronounced impact on the resulting cognitive map by conditioning, at a minimum, which elements of the environment are represented as well as in what form the relationships among them are encoded. This conclusion, in our opinion, suggests that anthropology has a special and unique place in the interdisciplinary research of human spatial cognition.

66 At the same time, some elements of this knowledge, for example knowing what certain places look like from above, will be rendered useless in the new configuration.

5 SOME ELEMENTS OF CONTENT OF SPATIAL COGNITION AMONG REINDEER HERDERS: THE REPRESENTATION OF MICROLANDSCAPE[67]

5.1 Microlandscape as an object of spatial cognition: The role of culture

Thus far our discussion of the interplay between culture and spatial cognition among the two reindeer herding groups has focused primarily on the structural aspects of spatial cognition. In this chapter, we shall broaden our discussion by focusing on some aspects of the *content* of cognitive maps among the nomads to reveal their cultural roots as well as their possible role in solving the cognitive tasks the nomads routinely face. Before proceeding to this, however, let us recall the general remarks on the content of cognitive maps made in chapter 3 to see how they can guide our discussion.

The elements of a cognitive map do not so much represent physical objects that exist in the environment, but rather the spatial aspects of an individual's knowledge and experience of the environment. From this alone it should be apparent that culture can have a huge impact on the content of cognitive maps because knowledge and experience of the environment depends to a large degree on culture. Culture can effectively inhibit one's experience of some elements of the environment and highlight the experience of other elements by, for example, encouraging a particular way of life or, as was described in the previous chapter, providing a particular cognitive system through which the experience is processed and stored. The absence of bogs in the cognitive maps of helicopter pilots described in the previous chapter is a good example of this. What is more important, however, is that cultural factors can determine particular kinds and forms of knowledge and experience that an individual gains from the environment (Ingold 2000). This suggests that representatives of different cultural groups can have different experiences of the same environment. These differences can involve, among other things, the spatial aspects of this experience.

These effects of culturally specific knowledge and experience on the content elements of cognitive maps can be expected to be especially pronounced in the case of what we referred to in chapter 3 as 2-D-S (two-dimensional surface) elements. Indeed, in comparison to the other types of elements on mental maps (points and lines), these elements are the least related to physical objects as such. They often represent such entities as parts of a city, neighbourhoods, places with high traffic volume at certain hours, areas characterised by a high crime rate, and so forth. In other words, these

67 This chapter is based in part on a chapter published in the collection *Nomadismus in der "Alten Welt"* (Istomin 2011).

elements tend to represent spatially related groups of places with certain properties or qualities. The principles on which such a grouping is based can vary greatly. We suggest that the degree and dimensions of this variation are not dissimilar to that of so-called folk categorisations – ways of sorting experience into conceptual categories – that have long fascinated cognitive anthropologists (see Atran 1990; D'Andrade 1990; Ellen 2003). Just as in the case of folk categorisations, it can be expected that some of the 2-D-S elements represented on cognitive maps correspond to areas that stand out of the surrounding environment in ways human cognition cannot ignore. In these cases, the principles of place grouping really "cut nature at its joints," as Plato once said (See Phaedrus 266a) and, therefore, one can expect such groupings to be present in all societies. For example, it can be expected that the cognitive map of any individual, irrespective of his or her cultural background, would contain 2-D-S elements representing lakes or mountains if such entities are present within the territory and if the individual has enough experience of that territory. Indeed, areas covered by water or containing difficult terrain have such distinct ecological characteristics and can make such a huge impact on human actions across different domains that distinguishing between them and other areas seems to be absolutely necessary to inform reasonably efficient spatial behaviour. Being able to make such distinctions is just as important for survival and successful action as the distinction between a cat and a tiger once referred to by Atran (1990). Similarly, it seems to us that the cognitive maps of all humans, independent of their culture, cannot fail to represent populated areas such as settlements or towns in apposition to sparsely populated or unpopulated areas. Indeed, given the inherent sociality of our species, being able to infer where conspecifics can or cannot be reliably met is too important for any strategy of spatial action to dispose with. These kinds of elements of cognitive maps can, in our opinion, be treated as analogous to the "natural kinds" of folk categorisations (See Atran 1990; Ellen 2003, 2006).

However, apart from these "natural" elements of spatial layouts, cognitive maps should be expected to include a large number of 2-D-S elements that are less universal. Many of these elements, we believe, reflect groupings of spatial locations based on qualities that make a difference only for certain domains or kinds of activities. Their presence on the cognitive maps of particular individuals, therefore, should depend on the degree to which these individuals are involved in the relevant kinds of activities on a day-to-day basis. Representing areas of a city that have a high volume of automobile traffic can be a good example of this. Indeed, one can expect these areas to be represented (as specific 2-D-S elements) on the cognitive map of a person who regularly drives through the city. On the other hand, such elements would probably be less salient for, or even missing from, the cognitive map of a person who does not own a car and normally walks through the city. Note that the reason here is not that the pedestrian cannot experience high traffic volume in the manner the helicopter pilots described in the previous chapter could not experience bogs. Indeed, the traffic on the streets can, in most cases, be easily observed and evaluated by a pedestrian walking on the sidewalk. The pedestrian could, therefore, represent areas with different traffic volumes on the basis of his or her experience. The reason that a driver represents these areas while a pedestrian does not is that the distinction between the areas is much more relevant for the activity of the former than the latter. The pedestrian simply does not experience the distinction in the same way and, therefore, does not pay attention to it.

As far as domains and kinds of everyday activities vary across cultures (and they certainly do), a significant degree of similarity of 2-D-S elements of certain classes among individuals from the same cultural group and, at the same time, a significant variation of these elements across cultural groups should be expected. One such class should include 2-D-S elements reflecting the form and quality of ground surface, that is the *microlandscape* of the territory represented. This includes, for example, representing meadows as specific parts of grassland, gullies as a distinct type of ravine, groves as specific areas in forests, and so on. Representatives of most, if not all, cultures do represent certain types of microlandscape elements: such representations are likely to have great practical importance for productive economic activities, which, in most societies, are based on certain forms of land use. On the other hand, we can expect great variation in the types of elements represented, depending on the exact forms of land use.

Our experiences among the two reindeer herding groups correspond well to these expectations. For us as well as for many immigrants from the south, even those who had lived in the region for dozens of years, the tundra landscape appears monotonous, with any of its parts looking very similar to all others. The reindeer herders, on the other hand, definitely perceived this landscape as consisting of a number of qualitatively different types of territory. The representation of areas in space in terms of these types features highly in their spatial cognition and makes up a very distinct and important class of 2-D-S elements on their cognitive maps. Furthermore, these elements represent one of the most important distinctive characteristics of the content (as opposed to the structure) of the cognitive maps of reindeer herders, especially when compared to those of other groups from the local population.

On the basis of these observations, we decided to make microlandscape elements the primary object of our inquiry into the impact that culture has on the content of spatial representations. This chapter presents and discusses the findings that resulted from this inquiry. We would like to stress, however, that we by no means believe that the impact of culture on the content of spatial representations is limited to microlandscape elements. We also do not think that microlandscape elements necessarily have some special status in the interplay between cognition and culture. Our choice was guided by theoretical considerations and empirical observations suggesting that the impact of culture could be pronounced in this aspect of spatial cognition. It can well be that the effects of culture are even more pronounced as far as other aspects of the content of cognitive maps are concerned.

5.2 Types of microlandscape among Nenets and Komi reindeer herders

The landscape categorisations used by the reindeer herders were recorded using the standard method for the collection of terminologies (see, e.g., Michrina and Richards 1996). Generally we started by picking out specific landscape terms either from our interviews with the reindeer herders or from conversations between reindeer herders that we witnessed. We would then ask the informants to explain the meanings of these terms. Quite often, especially if the herders were interviewed during a migration or while travelling across the tundra, the informants were able to point to areas we passed to illustrate the term. We would then try to use the concepts so explained to elicit other concepts of microlandscape belonging to the same taxonomic level. For example, if the informants explained that the concept we picked up from an interview referred to a special kind of flatland or bog, we would ask what other kinds of flatlands or bogs existed, and what they were called. After the initial lists of terms and concepts were compiled in this way, we started to make longer interviews with different informants, asking how the different types of territory referred to by these terms differed from each other. We also asked the informants to describe how areas belonging to these different types are distributed over the territory, and to point to some of them with their finger. Based on these interviews, we were able to make hypotheses about the essential qualities and distinctions that support the grouping of places into such areas. Finally, we checked these hypotheses during our trips in the tundra with the informants by pointing to different areas we passed through and naming the types to which, according to our hypotheses, they belonged. The informants could confirm or refute our statements. It should be noted here that our mastery of the Komi language has always been better than that of Nenets and, therefore, our methods worked much more effectively among Komi than among Nenets reindeer herders. Still, we managed to collect rather detailed lists of types of territories and their essential characteristics from both groups.

From these lists, two striking facts immediately emerged. First of all, most of the terms used by both of the groups to refer to the types of territories in the tundra were of Nenets origin. At first sight, this fact strongly supports the claim of Startsev (1926), subsequently reiterated by many later ethnographers of Komi (e.g., Konakov and Kotov 1991; Zherebtsov 1982), that the Komi have borrowed the whole complex of knowledge and concepts needed for living in the tundra from the Nenets (see chapter 1 for a more extended discussion). This claim, however, is refuted by the second fact that was immediately apparent: the landscape terminologies used by the two groups are very different. Despite the fact that most of the terms used by the Komi to refer to landscape elements are still present in the Nenets language (see table 2 below), these words, with only three exceptions, are not used by Nenets in their landscape terminologies. On the other hand, all but three terms the Nenets do use to refer to landscape elements are unknown to Komi. Furthermore, two out of the three terms used in both terminologies have rather different meanings for the Komi than they do for the Nenets.

What is much more significant, however, is that the underlying principles on which the microlandscape typologies are based differ considerably. For example, the Nenets differentiate three general types of landscape: *pödara* ("forest," i.e., large expanses covered by trees); *vy* ("tundra," i.e., land without trees); and *tanem*. The latter term refers to the flood plains of large rivers, which are usually either covered in bushes and small trees of the willow family or represent natural meadows and, indeed, stand out noticeably from forest and tundra environments. The Nenets say that *tanem* is neither forest nor tundra, but rather a special kind of land surrounded by forest or tundra.[68] The more specific microlandscape terms among the Nenets are based either on surface morphology or on vegetation type. Regarding surface morphology, the Nenets differentiate between hills (*sede*), long natural depressions or ravines between hills or on a flat surface (*liang*), and large expanses of flat areas resembling plains (*lapta*). Interestingly, the plains-like areas seem to be recognised as a specific type of landscape only in the tundra (*vy*). The informants never spoke of them in forested areas or within floodplains, and they could not point to *lapta*s in those areas, despite the fact that such areas appeared (to our eyes) to exist there. On the other hand, hills and ravines were apparently represented not only in the tundra, but also in forests and floodplains (the informants described them there and could point them out). Large slopes that often lead to large natural depressions, for example those leading down to river floodplains, are referred to as *iavsä*. Finally, natural furrows on such slopes, "which can be used for climbing up the slope with a reindeer sledge," are called *pinde*. This term is used irrespective of the size or the gradient of the slope, as well as of whether or not this slope is referred to as *iavsä*. Besides these, we managed to record two terms reflecting the physical quality of the surface. These are *ium* (variant – *iumbar*), referring to polygonal tundra,[69] and *ilä ia* (lit. "living land" or "alive land"), which refers to swampy territories "where the ground moves under your feet as if it were alive." Both were described as special kinds of *lapta* (tundra plains) and were said to be absent in the forest or in *tanem*.

Concerning the types of territory defined by vegetation, we should first mention two terms – *paiu' hoi* and *nero' sä* – that were rather frequently and regularly used by the Nenets informants. These terms refer to large areas covered by one or the other of two species of bushy willow (*nero* and *paiu*) that are frequently encountered in the tun-

68 Interestingly, some Russian ecologists are of the same opinion. Chernov, in his classic work on tundra ecology (1980), argues that flood plains should be analytically differentiated from other large ecological zones (such as tundra, taiga, steppe, etc.) and treated as "interzonal landscapes" having special characteristics of their own across all the zones.

69 This natural phenomenon is caused by the large masses of permafrost situated close to the surface and producing the so-called cold boiling of the soil in winter. This results in a very distinctive surface consisting of rectilinear areas (polygons) 8–13 sq. meters in size separated by narrow (10–20 cm. wide) but rather deep clefts.

dra. *Paiu' hoi* refers to a place where "the land is covered by *paiu* willow all the way to the horizon" (from *paiu*, gen. and *hoi* – "horizon"). Similarly, *nero' sä* refers to the area covered by *nero* bushes (from *nero*, gen. and *sä* – "surface"). The territories referred to by these terms were viewed by the informants as particular types of tundra (*vy*), and they were clearly independent of surface morphology. Informants told us that *paiu' hoi* and *nero' sä* are "large areas in the tundra (*vy*), where there can also be hills (*sede*) and depressions (*liang*), but they are all covered by willow." Besides that, the vegetation could be, and often was, referred to in composite expressions referring to the tundra plains (*lapta*). Thus, the informants could speak of *pünk' lapta* (tundra plains covered in dwarf birch (*pünk*, gen.)); *nädo' lapta* (tundra plains covered in lichen), and so forth. Such composite expressions, however, did not represent regular terms and the informants did not mention them in response to our inquiries about taxonomic kinds (e.g., "What [other] kinds of *lapta* are there?"), whereas other composite expressions such as *ilä ia* were mentioned. This suggests that the Nenets informants think about such expressions as referring not to special types of *lapta*, but rather to its non-categorical characteristics. On the other hand, it seems to us that these characteristics do produce separate 2-D-S elements on the cognitive maps of the Nenets reindeer herders. During interviews they could usually point out *punk lapta, nädo lapta*, etc., as separate areas. It is also significant that the Nenets herders very rarely referred to other types of surfaces on the basis of vegetation. They did not normally speak of, for example, *pünk liang*s (which would be natural depressions covered by dwarf birch) or *nädo' sede*s (hills covered in lichen). Although they understood such terms when we produced them and used them to refer to concrete hills or depressions, many of them remarked that people "do not normally say that – they would simply say '*sede*' or '*sede* where lichen grows' if they really need to describe this." Furthermore, although most Nenets herders could point to a number of hills situated in the regions where interviews were performed (as well as, sometimes, in adjacent regions), only a few of them could point to hills with specific vegetation. This represented a striking contrast to parts of tundra plains (*lapta*) with a specific vegetation – most reindeer herders could point to a number of such areas. This suggests that Nenets herders normally represent the spatial distribution of vegetation over tundra plains but not over other types of territory as producing specific areas (2-D-S elements). The types of microlandscapes detected among the Nenets are summarised in table 1.

Level of typology	Term	Description
General types of landscape		
	vy	tundra
	pödara	forest (taiga)

	tanem	floodplain
More specific types of microlandcsape		
Basic		
	sede	hill
	liang	long natural depression
	iavsä	larger slope of a hill or natural depression
	pinde	natural furrow on a larger slope
	lapta	tundra plain
Specific kinds of tundra plains (lapta)		
	ium	polygonal tundra
	ilä iä	swampy area
	pünk lapta, nädo lapta, etc.	areas of tundra plains covered by a certain type of vegetation (dwarf birch, lichen, etc.)
other		
	paiu' hoi, nero' sä	large tundra territories covered by willow stands

Table 1. Types of microlandscapes that are likely to make up spatial representations on the cognitive maps of Nenets reindeer herding nomads.

In contrast to the Nenets, Komi reindeer herders distinguish between only two general types of landscape: forest (*vör*) and tundra (*tundra*). The latter term, which is believed to be of Finnish origin and is used in many European languages, probably entered the Komi language via Russian. Most of the more specific landscape-based types of territory are differentiated on the basis of surface morphology *and* vegetation, or vegetation only. These types are, therefore, specific to either the tundra or the forest. The exceptions are the terms *mylk* (hill) and *musur* (hill range): they reflect surface morphology only and are, therefore, used to refer to particular types of territory in both the tundra and the forest. Significantly, these terms are not borrowed: they are purely Komi terms that are also used by groups of sedentary Komi in the south (see chapter 1).[70] Most other concepts used for tundra landscape are designated by Nenets

70 In the tundra, natural depressions of any kind can be referred to as *göp* (pit). However, since this

words, sometimes with the addition of the Nenets productive suffixes *-ei* (the suffix of collective or abstract nouns, like *-schaft* in German or *-ship* in English) or *-ko/-ku* (affectionate diminutive suffix, like *–schen/-lein* in German).[71] Thus, an elevation with a dry top covered in lichen is referred to as *iarei* (cf. Nenets *iar* – "sand"). An elevation covered by grass is called *veret'a*. A humid tundra plain covered in grass and sedges is referred to as *sada* (cf. Nenets *sada* – "puddle"). A boggy tundra covered in sedges and with water blinking on its surface is called *saduku* (Nenets *sada* + *-ku*). A swampy tundra not covered in sedges, but "where the surface goes up and down under your feet and mud sometimes comes out from beneath" is called *tabei* (cf. Nenets *tab* ("mud") and the Nenets explanation for *ila ia* above). A relatively dry, flat area in the tundra sparsely covered by small bushes (not willow!) and lichen is called *lapta* (cf. the similar term in the Nenets landscape terminology meaning simply "tundra plain"). An area covered by willow (i.e., a willow stand) is called *röshsha* (probably from the Russian *roshscha* – "grove"), irrespective of its relation to the surface morphology. An area densely covered by dwarf birch or other small bushes is called *pacha* (etymology unknown to us), also irrespective of its relation to morphology. Finally, a steep stony or sandy hill with very sparse or no vegetation is called *lotsovku* (etymol-

An example of *pinde*. Photo: Mark Dwyer.

latter term is applied to any kind of depression, including artificial excavations in the ground, hollows made in the snow, and even cavities made on a piece of wood, we are not sure that it represents a microlandscape concept.

71 Interestingly, these suffixes are rarely used with the same words by Nenets themselves.

ogy is unclear, but note the Nenets suffix *-ku*). The same hill with natural furrows on its slopes is called *pindei* (cf. *pinde* in Nenets landscape terminology, meaning a natural furrow on a slope leading to a large depression).

The landscape concepts used in the forest zone include, apart from the already mentioned cross-zonal concepts *mylk* and *musur*, the categories *choi* (used also by sedentary Komi to designate ravines or larger natural depressions, cf. the Nenets term *liang*), meaning a treeless ravine or larger depression in the forest zone, and *nur*, meaning a bog without trees. Relatively dry areas covered by sparse forest are called *iag*. All other areas covered by trees are called *vör* (forest), the term that is also used for the ecological zone as a whole. Finally, dry areas without trees inside the forest zone are called *tundra tor* (*tor* – Komi: "part," "piece"), while small patches of forest in the tundra are called *vör di* (*di* – Komi: "island"). Note that these forest zone terms are not borrowed: all of them (with the exceptions of the composite terms *tundra tor* and *vör di*) are of Komi origin and are also used by the sedentary southern Komi. The Komi landscape concepts with the etymology of the terms used to refer to them are summarised in table 2.

Level of typology	Terms	Etymology R. – Russian; N. – Nenets; K. – Komi	Description
Main types of landscape			
	tundra	R., originally borrowed from Finnish	tundra
	vör	K.	forest (taiga)
Microlandscape-based types of territory in both types of landscape			
	mylk	K.	hill
	musur	K.	hill range
Microlandscape-based types of territory in the tundra			

	iarey	N. *iar* – "sand"	a long dry hill with lichen on its top
	veret'a	N. *verta* – "grass"	a long hill with grass on its top
	sada	N. *sada* – "puddle"	flat wet area covered in grass and sedge
	saduku	N. *saduku* – "little puddle"	flat wet area covered in sedge with water on the surface
	tabey	N. *tab* – "mud"	swampy tundra
	lapta	N. *lapta* – "tundra plain"	flat dry area covered in lichen and sparse bushes
	röshsha	R. *roshscha* – "grove"	area covered in willow
	pacha	?	area covered in dwarf birch
	lotsovku	?	steep hill without vegetation
	pindei	N. *pinde* – "furrow on a slope"	steep hill without vegetation but with natural furrows on its slopes (possible to climb up with a sledge)
Microlandscape-based types of territory in forest			
	choi	K.	depression without trees
	nur	K.	bog without trees
	iag	K.	sparse forest

Table 2. Terms for types of microlandscape that are likely to make up spatial representations on the cognitive maps of Komi reindeer herding nomads.

To summarise, the reviews of Nenets and Komi typologies show that while Nenets structure the microlandscape in terms of either morphology of surface or vegetation, Komi generally do this by combining the two. This important difference in the principles underlying the typologies exists despite the long historical interaction between the two groups, the similarity of the categorisations they apply in many other important domains, and the peculiar fact that Komi use Nenets words to designate most of their landscape concepts in the tundra zone. Importantly, the basic principles described are not applied entirely consistently by both groups. Thus (pseudo)types such as *punk' lapta* and *nädo' lapta* among the Nenets seem to group places in the tundra on the basis of both surface morphology (tundra plain, *lapta*) and vegetation (dwarf birch, lichen, etc.) Similarly, the Komi typology includes several concepts, such as *mylk* and *musur,* that refer to territory on the basis of surface morphology alone. In the next section of this chapter, we will attempt to explain why these differences exist and what they mean.

5.3 The land for herding and the space for traveling

The microlandscape-related 2-D-S elements of cognitive maps can be expected to reflect groupings of spatial locations based on qualities that are particularly relevant to specific kinds of land-use activities. In other words, such qualities can be expected to be exactly those that are especially useful for guiding the inferences and actions related to particular forms and methods of land use. The first and most obvious place to look for an explanation of the differences in the microlandcsape-related 2-D-S elements between the Nenets and the Komi is, of course, the particularities of the reindeer herding practices of the two groups. Details related to these practices have been discussed in chapter 2, where we concluded that the cornerstone of the reindeer herding practices of the two groups, as well as arguably of any practice of nomadic pastoralism, seems to be a particular interplay between the behaviour of the animals and the actions of the herders that ideally results in the herders maintaining control over their herds and minimising the chances of loss of livestock. This interplay, in turn, is based on the ability of the herders to effectively interpret the behaviour of the animals and skillfully react to it in such a way as to prevent the loss of control over the animals.

As was also discussed in chapter 2, Komi and Nenets have developed different ways to achieve this interplay. Komi reindeer herds are under the constant supervision of a duty herder throughout the spring, summer, and early autumn. The primary way to achieve control over the herd among the Komi is through skillful maneuvering of the herd over the pastureland by carefully directing the herd's movement. Using his knowledge about the factors influencing reindeer behaviour, the herder regulates the speed of the animals' movement and the general dispersion of the herd over the pastureland. The two most important factors determining the impact of the landscape on reindeer behaviour are the morphology of the surface and the kind and quantity of reindeer forage. Furthermore, these factors are among the main predictors of reindeer behaviour, along with weather conditions (temperature, strength and direction of wind), depth and quality of snow cover, and (in summer) mosquitoes and other insects that harass the animals. Furthermore, the two landscape-related factors affect the herd simultaneously, and the Komi herders attempt to manipulate exactly these factors through their skillful maneuvering. It makes perfect sense, therefore, that Komi mark out areas in the tundra (and represent them as 2-D-S elements on their cognitive maps) with reference to both surface morphology and vegetation as basic criteria.

Indeed, this delineation of landscape immediately structures it into "the land for herding," which for the Komi primarily means places where they are best able to direct the reindeer in a way that ensures the herd's coherence and productive grazing. Let us suppose that each area marked out in this way is associated by a Komi herder with a particular range of herd behaviour that can be expected under different weather conditions and other relevant factors. Let us further suppose that these areas differ in the related risks to the herder's success in maintaining cohesion of and control over the herd. Using his representation of the spatial distribution of such areas, the herder can infer a strategy of maneuvering that would minimise these risks given the other factors present. For example, on a warm summer day when the herd is harassed by mosquitoes, the reindeer usually disperse in search of protection from the insects (see chapter 2). Under these conditions, the herder can choose to maneuver the herd in such a way as to spend the late morning and/or early evening hours – the periods when mosquitoes are usually particularly abundant – on the slopes of a *vereta* or a *iarei* (long hill covered with grass and lichen respectively). In these areas, the wind blows the mosquitoes away, providing some relief for the animals and, furthermore, the animals are likely to find enough forage there to graze calmly for several four-hour activity/passivity periods (this is more likely the case for a *vereta* than a *iarei*). This reduces the risk of losing control over the herd. On the other hand, this decision potentially increases the risk of the herd splitting up, as some animals might run over the top of the hill and lose contact with the rest of the herd. The herder could reason that during the periods of the day when mosquitoes are most abundant, this risk is offset by the reduced chances of losing control over the herd. However, his likely strategy of maneuvering would most probably include taking the herd down to adjacent *lapta*s when the mosquitoes are fewer in number. Finally, Komi reindeer herders believe that *pacha*s – flat areas covered in dwarf birch – represent the most valuable bits of pastureland throughout the summer because, under normal conditions, reindeer herds move over them very slowly and stay remarkably coherent. If such areas are present within the territory that is available to the herder for his twenty-four-hour shift, it would behoove him not to bring his herd there during the day, but to reserve such pastures for the night grazing. This would allow him to rest and maybe even to sleep for several hours after the day's work and, if the nights are already dark, greatly reduce the risk that the herd might escape from his control during the period of poor visibility.

Even the apparent exceptions from the basic principles of microlandscape typologisation among Komi can be easily explained from the viewpoint of their importance for maneuvering the herds. For example, the hills and hill ranges can affect the coherence of the herd regardless of the vegetation. Similarly, the bush (willow) stands are not suitable areas for maneuvering the herd regardless of their landscape morphology and the amount of reindeer forage they might contain.[72] Indeed, willow stands are usually rather dense and can reach a height of one and a half meters, rendering them virtually impossible to traverse on a reindeer sledge. Moreover, a herd of reindeer hidden in a willow stand is very difficult to observe and round up. Thus it makes sense that Komi reindeer herders would represent *mylk* and *musur* (which are not dependent on vegetation) and *röshsha* (which is not dependent on landscape morphology) as distinct 2-D-S elements on their cognitive maps, because these concepts do in fact emphasise distinctions that are important for reindeer pasturing. Furthermore, the distinction between two kinds of steep hills without vegetation – *lotsovku* and *pindei* – which emphasises the presence or absence of natural furrows on their slopes (which can be used to ascend on a reindeer sledge), can also be a part of this picture: Komi herders usually use such hills as observation points to oversee the herd, evaluate its behaviour, and plan future actions.

The herding techniques of the Nenets herders are substantially different from those of the Komi. We mentioned in chapter 2 that Nenets normally do not constantly oversee their herds. The herd is allowed to disperse and graze without oversight most of the time. Rather, the herders round up the herd once or twice a day, bring it to the *tandara* (the area immediately adjacent to the camp), and then "push it out" from there in a certain direction to disperse again. Deciding on the direction in which to "push out" the herd, and then tracking the dispersed groups of reindeer to round the herd up again are, of course, non-trivial cognitive tasks that are likely to involve a great deal of knowledge and understanding of reindeer behaviour. As the description in chapter 2 suggests, to perform these tasks the Nenets skillfully evaluate the weather conditions (particularly temperature and wind direction), the needs of the reindeer at a given time of the year, and the landscape morphology and vegetation.

We can suppose, then, that despite the differences in herding techniques, the body of knowledge relating reindeer behaviour to a set of biotic and abiotic factors, including the microlandscape, is likely to be similar among the Komi and the Nenets. There is, however, a difference in the ways the two groups apply this knowledge: the Komi use it to manage the skillful maneuvering of the herd, while the Nenets use it to predict the whereabouts of scattered groups of reindeer after hours of unobserved grazing and dispersal. In other words, while the Komi have to predict the immediate effect

---

72 It should be noted that young willow leaves provide very nutritious forage for reindeer, so both the amount and quality of reindeer forage contained in willow stands can be quite high, especially in the spring.

of a certain constellation of relevant factors on reindeer behaviour, the Nenets routinely face the task of determining the relatively long-term cumulative effects of such factors, taking into account the probable change in the factors over time. Since the changes in any particular factor are only remotely (if at all) correlated with changes in other factors and, furthermore, particular factors are likely to affect, in the long run, certain aspects of reindeer behaviour more than others, it makes perfect sense to approach the task by considering each relevant factor separately and adding its effect to the cumulative effect of other factors already considered. The concrete cognitive mechanisms likely to be employed by the herders in order to do this will be discussed in more detail in chapters 6 and 7. Here it is important to note that the different aspects of microlandscape such as vegetation and surface morphology are among those factors that can be usefully considered separately rather than together. Indeed, although both of these aspects simultaneously participate in generating the specific pattern of behaviour exhibited by a reindeer herd at any given moment, their long-term effects can be dissociated. As the example of Nenets pasturing in chapter 2 suggests, the long-term effect of vegetation, as perceived by the herders, relates to the speed of the herd's movement (and, hence, the distance the herd is likely to cover in a given time) more than the direction of movement. In contrast, surface morphology affects the direction of movement and the likelihood that the herd will splinter off into groups, and only to a much lesser degree does it affect the speed of the herd's movement. Indeed, it seems likely that a herd of reindeer will follow hill slopes and natural depressions independent of the vegetation it finds there, while the vegetation is likely to determine the speed of the herd's movement independent of the surface morphology of the landscape.[73]

Taking these effects into account, it again makes perfect sense that Nenets represent such aspects of microlandscape as surface morphology and vegetation as separate 2-D-S elements. Furthermore, it also makes sense that the areas with different types of vegetation are represented as separate elements only in the flat tundra (*lapta*). Indeed, since western Siberian tundra is relatively flat, we can reasonably expect that the reindeer would spend considerably more time grazing on (and moving over) *laptas* than hills or other areas with complex surface morphology. Therefore, the distance they would cover would be affected mainly by the vegetation they would find in these areas. The areas with complex landscape morphology, on the other hand, can significantly alter the direction of reindeer movement, while the vegetation in those areas can have a significant immediate effect on the speed of movement. However, its overall effect on the distance travelled by the herd before the next roundup is, most probably, relatively insignificant due to the fact that such areas make up a rather small proportion of the territory through which the animals travel. For this reason the Nen-

---

73 There is also a correlation between type of vegetation and surface morphology that is particularly robust in the tundra (Chernov 1980). It is quite likely that the microlandscape typologisation of the Komi makes some use of this correlation.

ets herders can probably ignore this effect when predicting the location of the herd (or its different parts) for the next roundup.

The way Nenets reindeer herders can achieve such a prediction, therefore, can be sketched out as follows. Suppose that the herd was "pushed out" from the *tandara* toward a particular flat area (*lapta*) adjacent to it. Using the representation of this area and of the spatial distribution of vegetation over it (encoded on their cognitive maps as 2-D-S elements such as *pünk' lapta, nädo' lapta,* etc.), and taking into account other relevant factors (wind, temperature, etc.), the herders can infer how much time it will take for the animals to go through this area before they will encounter an area with a more complex surface morphology, let's say a hill, a ravine, or a stream.[74] Then, using the representation of that area, they can infer how the direction of the herd's movement would change there, and to what new *lapta* this would bring the animals.[75]

Then, a new inference of the distance the animals are likely to cover over this new *lapta* would be made and, if this inference indicated that the animals would likely reach another area of complex landscape morphology, yet another new inference about their change of direction would be attempted. The result of this process would be a rough estimation of where their herd is likely to go after pushing out from the *tandara*, and where it is likely to be by the time of the next roundup. Although we do not have hard empirical evidence that this description is valid, it does, in our opinion, correspond well to the accounts of reindeer herders themselves about how they predict the movement of their herds (some of which are given in chapter 2). In any case, it represents a good working hypothesis of how the typologisation of microlandscape observed among the Nenets is related to their reindeer herding practice.

Another domain of everyday activity for which the distinctions supporting the typologisation of microlandcsape among the Nenets can potentially be useful is travel by reindeer sledge. Indeed, in the treeless tundra, such travel, in most cases, is not affected by vegetation. An important exception is willow stands, which, as has already been mentioned in relation to maneuvering the herds among Komi, cannot be easily traversed on a reindeer sledge. In contrast, surface morphology is of great importance for traveling: finding one's way in the tundra on a reindeer sledge involves frequent climbing up steep slopes of hills and ravines, and, in the winter, guessing where snow is harder or shallower (again, depending on how the place is situated in relation to hills and ravines, which are the main factors influencing the accumulation of snow). On a comparatively flat surface, travel is influenced by swamps and the areas of polygonal tundra – the places where sledge skies and reindeer legs are most

74 Of course, the herders can also infer that the animals would not reach any such area before the time of the next roundup. In this case, this inference alone would give them the likely position of the herd.

75 They can also infer how likely it is that the herd would split into groups there and in what directions (toward which flat areas) these groups would likely go.

commonly broken. It is exactly these landscape particularities that are emphasised in the Nenets typologisation of microlandcsape, which carefully differentiates between kinds of slopes and divides the flat tundra into dry places, swamps, and polygonal tundra. Therefore, while the Komi typologisation of microlandcsape structures it into the land for herding, the Nenets typologisation, it appears, structures it into both the land for herding and the space for traveling.

5.4 Conclusions: Culture, reindeer herding, and spatial cognition

The empirical material and discussion presented in this and the two previous chapters has indicated a number of particularities of spatial cognition among the two groups of nomadic reindeer herders. These particularities make their spatial representations (cognitive maps) different in terms of both structure and content from those of many settled groups. It is also significant that important differences in the structure and content of spatial representations were also observed between the two groups of reindeer herders themselves. Now is a good time to consider how one can make sense of these particularities from the viewpoint of the existing approaches to the interplay between culture and cognition that were briefly sketched out in chapter 1, as well as from the viewpoint of the cognitive tasks the reindeer herders from the two groups face on a daily basis.

Many of the particularities observed can be comfortably placed into the category of "cultural models" (including scripts) as they are understood by D'Andrade (D'Andrade 1990; D'Andrade and Strauss 1992) and others (e.g., Holland and Quinn 1987; Shore 1998; Strauss and Quinn 1998). This concerns practices related to spatial navigation such as position fixing, course taking, and course keeping (discussed in chapter 4), as well as the selection of qualities of microlandcape that form the basis for the grouping of places into 2-D-S areas (as discussed in this chapter). Indeed, these aspects of spatial cognition can be viewed as prescriptions or recipes that specify the deployment of particular cognitive "tools" or processes (such as categorisation routines or cognitive maps in general) in order to facilitate the effective execution of particular cognitive tasks. Some of these tasks are rather specific to reindeer herding (e.g., inferring an optimal strategy of maneuvering the herd or predicting its spatial location after hours of unobserved grazing), others are less so (e.g., finding their way toward a particular location in space). In any case, however, these are the tasks that the herders face on a daily basis by virtue of their economic activities (reindeer herding) and their nomadic way of life. The difference in such "cultural models" between the two reindeer herding groups can be partly explained by the differences in the way they practice reindeer herding and, probably, the way they nomadise (following established migration routes vs. the absence of such routes). However, some rather interesting differences in these models, particularly those related to position fixing and course taking, seem to resist such an explanation. Since we are hesitant to invoke the notion of "tradition," which, in our opinion, is highly uninformative and represents a kind of all-encompassing "black box" not unlike "God's will," we have to accept that no satisfactory explanation of these differences is known to us at the moment.

The remaining particularities observed among the reindeer herders in the domain of spatial cognition are more interesting. These particularities seem to represent something more fundamental than simple recipes or rules of thumb that specify deployment of particular cognitive "tools." Rather, they seem to be related to the design of the "tools" themselves, which brings us to the "cognitive toolkit" approach to the study of relations between culture and cognition (see chapter 1). Let us recall that, according to this approach, cultures not only specify "the tools of choice" for a particular cognitive problem, but can also "construct composite cognitive tools out of the basic universal toolkit, thereby performing acts of elaborate cognitive engineering" (Nisbett et al. 2001: 306). The empirical material presented in the previous chapters seems to suggest that, as far as their structure rather than their content are concerned, the cognitive maps of reindeer herders can be interpreted as examples of such composite cognitive tools. Indeed, as was demonstrated in chapter 3, one of the particularities of spatial representations among reindeer herders is their heavy reliance on the survey type of representation (mental maps or survey knowledge) as opposed to the representation of routes between spatial locations. It seems like the reindeer herders possess survey knowledge of unusually extensive territories and, furthermore, their spatial representations that take this form include an unusually large number of locations. The only "sedentary" group in our study whose representatives seem to have mental maps of comparably extensive territories (or even larger ones) is helicopter pilots, but their mental maps still do not represent as many different spatial locations as those of the reindeer herders. The difference between the relative roles mental maps and route knowledge play in the spatial cognition of reindeer herders and sedentary groups is, of course, one of degree rather than of kind. This is exactly what psychocultural universalism as a general theoretical position would predict: the fundamental psychological processes, to which the ability to form spatial representations in the form of mental maps and to memorise routes as temporal sequences of environmental features most probably belong, are universal across cultures and societies. However, the difference we have observed in the relative salience of these two types of spatial representation is evidence of the ability of culture to influence the very *structure* of representing space mentally rather than simply the content of representation or its deployment. Here we meet a genuine example of culture constructing a cognitive "tool" out of the basic universal toolkit.

In chapter 3 we already started to develop an argument that this difference in the structure of mental representations between the herders and most of the settled population can also be accounted for by a particular class of cognitive tasks they frequently face: those of finding their way in the tundra and taiga under conditions that place serious limitations on visual perception, such as during (polar) nights and in fog and snowstorms. Under these conditions, it can be difficult or even impossible to perceive the sequences of environmental features that specify memorised routes. On the other hand, given the pervasiveness of such conditions in the environment the herders inhabit, as well as the role traveling plays in their lives, the tasks of navigation under these conditions cannot be avoided. The extensive use of abstract representations of Euclidian spatial relations between locations, which do not depend to the same degree on perceptible environmental features for their use in navigation, seems to be an obvious solution. In this chapter we have discussed other cognitive tasks that are frequent in the lives of the herders and that require mental maps to be successfully carried out. These are the tasks of devising optimal strategies of maneuvering the herd (among Komi) and predicting the whereabouts of reindeer (among Nenets), both of which seem to be heavily dependent on the representation of the spatial distribution of microlandscapes. We argue that these microlandscapes have to be represented as 2-D-S elements, which can be integrated into a cognitive map only by means of survey knowledge. Indeed, knowing only routes in the tundra, even if this knowledge also involves knowing the surface morphology and vegetation along those routes, is hardly enough to complete these tasks.

Other particularities of spatial cognition among the reindeer herders that can also be examples of culture's engineering of cognitive tools are related to the structure of cognitive maps or survey knowledge. This includes the specific mode of representation of distances between locations in *nedalava*s among Nenets, as well as the regionalised character of their mental maps (among Komi, mental maps do not seem to be regionalized). We have already argued that the former particularity can be, most probably, treated as a consequence of an "extended" (e.g., Clark 2010) or "distributed" (e.g., Hutchins 1995) cognitive process, in which the Nenets use their transport animals – which have a particular non-random cycle of run and rest – as a part of their cognitive routine for learning and memorising information about distances, as well as for retrieving and using this information for practical purposes. Given that these animals have been present in virtually all situations in which Nenets are learning a novel environment or navigating a familiar one – a fact that is, of course, directly related to the culture, economy, and way of life of these people – this extended cognitive process has been a fairly effective solution to the task of learning and using information about distances. Our material demonstrates, however, that the reliance on this extended process poses significant problems once the Nenets face the rather novel task of traveling in the tundra by snowmobile. Since this task is becoming more and more pervasive, we would expect a modification of the extended cognitive process or its replacement by a new cognitive mechanism in the future, which will lead to changes in the structure of mental maps among Nenets. It should be noted, however, that the example of *nedalava*s as the basis for representation of distance among Nenets illustrates an interesting and thus far largely neglected way in which culture can construct cognitive processes: it can do so by providing particular objects, configurations of external features, and/or situations that can become a part of extended cognitive routines.

Finally, the regionalised character of mental maps among Nenets can be related, as we tried to demonstrate in chapter 4, to the particular practices of position fixing and course taking among this group of herders. Indeed, it seems to us that these practices would have simply failed to work if the mental maps of Nenets did not have such a regionalised structure. This makes it possible to suggest that the regionalisation of the mental maps has been caused by the cultural scripts that define their usage. In other words, it brings to our attention another interesting possibility: that culture can construct cognitive "tools" by providing models of their application. Unfortunately, since we do not have a convincing account that would explain the origin of the cultural models in this particular case, we cannot judge how valid this suggestion is. Indeed, it could just as well be the case that the regional structure of mental maps among Nenets gave rise to the particular practices of position fixing and course taking. More research is needed to prove or disprove any of these possibilities.

The empirical material on spatial cognition among the two groups of reindeer herders in comparison to that of sedentary groups suggests that culture, understood broadly to include productive practices and ways of life, can influence cognition on several levels and in several ways. It can create models or scripts specifying the ways in which cognitive processes are deployed. It can also affect or engineer the cognitive processes themselves by constructing new and culturally specific cognitive routines out of simpler and, most probably, universal processing components. This can be achieved through both direct and indirect impacts of culture upon cognitive processing. The indirect impact can, for example, consist in creating particular milieu (material as well as social) for an individual; in this milieu, there can be particular features and configurations that can become a part of extended cognitive routines impossible in other milieus. These extended routines would, therefore, be culturally specific. Another way in which culture can indirectly influence cognitive processes is through creating cultural models that would demand specific cognitive tools for their realisation.

The principal way in which culture impacts cognition, however, is most likely by creating everyday situations that pose particular and non-trivial cognitive tasks for individuals. Most of the culturally specific aspects of spatial cognition among reindeer herders can be viewed as specialised mechanisms developed for performing such tasks, particularly those of spatial navigation under specific and complicated conditions, and analysing and predicting reindeer behaviour. The latter task, however, is likely to have a pronounced impact far beyond the sphere of spatial cognition. The second part of this book, therefore, discusses the reindeer herders' approach to this task in greater detail.

6 REINDEER BEHAVIOUR AS THE PRODUCT OF DYNAMIC MUTUAL ADAPTATION BETWEEN HERDERS AND ANIMALS[76]

6.1 Reindeer behaviour and the problem of animal culture

A good place to start our discussion of the cognitive mechanisms and routines that the reindeer herders can use to predict and make sense of reindeer behaviour is to ask what this behaviour actually represents and how it is formed. Although at first glance it may appear that this question is related to biology and animal ethology rather than to anthropology or cognitive science, we hope to demonstrate in this chapter that this is only partly true. We will try to prove that the patterns of reindeer behaviour are shaped, to a very significant degree, by the actions of reindeer herders. This means that the behaviour of reindeer is, in an important sense, the product of the culture of their masters. This, in its own turn, puts the problem of interpreting and predicting it into an entirely new light.

In chapter 2 we noted that the tradition of research on nomadic pastoralism in anthropology has long been plagued by a kind of naive anthropocentrism: the failure to recognise that the patterns of animal behaviour play a role in the functioning of nomadic pastoralist systems. In many classic models of nomadic pastoralism (e.g., Barth 1960, 1961; Gulliver 1975; Spooner 1972) animals are assumed to have needs but no behaviour of their own to satisfy them. The herders, using their knowledge and will, satisfy these needs by moving the animals from one place to another, grouping and regrouping them, and so forth. The empirical evidence against this view is abundant in the anthropological literature (e.g., Coppock et al. 1986; Erdenebaatar 1996; Swidler 1972; Turner 1999), and it eventually led to a more balanced approach to nomadic-pastoralist systems, according to which the animals do have important behavioural patterns of their own, and the interaction between the animals and the herders represents an important part of nomadic pastoralism. In the analysis of reindeer herding nomadism, this approach is evident, for example, in the work of Robert Paine (1994b) and Hugh Beach (1981, 1990). In chapter 2, we have developed this approach into an explanatory model on the basis of our data on two groups of nomadic reindeer herders.

We wish to argue, however, that although this more balanced approach to nomadic-pastoralist systems represents an important step toward cleansing nomadic-pastoralist studies of their anthropocentric bias, it still remains biased to a certain

76 This chapter is based in part on a research article published in *Human Ecology* (Istomin and Dwyer 2010).

degree itself. Indeed, this approach explicitly focuses on the ways in which nomads take into account and react to the behaviour of their animals, as well as on the role this process plays in shaping their migration patterns, the size of their herds, the patterns of pasturing, and other factors related to nomadic pastoralism. However, it largely ignores the possibility that similar processes exist on the side of the animal counterparts: that the reaction of the animals to the behaviour of their human masters can itself shape the behavioural patterns of the herds and the individual animals within them. It would appear that this approach implicitly assumes that animal behaviour is genetically prescribed and its patterns are biologically given and therefore constant, at least in the short and medium terms, until they are changed by the process of natural selection. Thus, although humans can modify their behaviour in response to that of the animals, the possibility that the animal counterparts can likewise modify their behaviour is precluded. This assumption is, however, quite unwarranted from the viewpoint of animal ethology and animal psychology.

As early as 1925, a short article provocatively titled "Have subhuman animals culture?" (Hart and Pantzer 1925) was published in the *American Journal of Sociology*. This article, despite being rather naïve by modern standards and therefore rarely referred to at present, initiated a long-lasting debate in the biological sciences concerning the formation and evolution of animal behavioural patterns. Based on zoological observations they collected, Hart and Pantzer argued that innate – that is, genetically inherited – behavioural patterns do not account for all or even a significant part of animal behaviour. Rather, the authors assert that "[b]ehavioral patterns in general might be classified into three groups: 1) instinctive, 2) sub-socially habitual, and 3) cultural" (Hart and Pantzer 1925, 705). The first group includes behavioural patterns which are, indeed, genetically inherited and, therefore, "develop full-fledged without the aid of trial and error, imitation or tuition if a suitable environment is provided" (Hart and Pantzer 1925:705). In contrast, "the sub-socially habitual group includes action patterns which, while depending upon innate capacities, are conditioned in their form by the varying nature of the physical environment and by learning through trial and error" (Hart and Pantzer 1925:705). Finally, the group of cultural patterns "includes all behavioural patterns acquired through social contacts, i.e., by imitation or tuition" (Hart and Pantzer 1925:705). Special attention in the article is paid to the last group of behavioural patterns. The authors convincingly demonstrate that these patterns play an important, albeit rather neglected, role in the formation of animal behaviour, and contend that they are not dissimilar to manifestations of human culture if culture is broadly understood as any socially transmitted knowledge and behaviour. From this point of view, the authors argue, the behaviour of semi-domesticated animals can be best understood in terms of cultural transmission:

> The definition [of cultural behavioural patterns] does not stipulate that the behaviour pattern need have originated in the species which acquires it. The domestication of animals consists in transmitting to them certain culture complexes. The fire-engine horse,

> in his response to the ringing of the alarm, in his habitual adjustments to the harness and to the rest of the apparatus, in his participation in the excitement of the fire run – in his whole adaptation to the situation – displays his adaptation of a culture complex. When, moreover, a new horse in being broken in is placed between two old horses, the animals themselves are participating in the transmission of the culture. (Hart and Pantzer 1925: 706)

Later studies in animal ethology have generally confirmed the intuitions of Hart and Panzer. For example, it has been convincingly demonstrated that the behaviour of the majority of complex animals is quite flexible and indeed cannot be regarded as inherited or "innate" in the traditional sense of these words (see Heyes and Galef 1996). Furthermore, formal mathematical models of the evolutionary process predict that complicated behavioural complexes completely programmed by genes are probable only among the inhabitants of a relatively stable environment (Boyd and Richerson 1985; Laland et al. 1996). Indeed, biological selection, the only known process of genetic adaptation to the environment, is a rather long process and, therefore, it cannot be used by animals to adapt to rapid and frequent changes (Boyd and Richerson 1985; Laland et al. 1996; Lefebvre and Giraldeau 1996). For these reasons greater reliance on mechanisms of adaptation via experience, including those that are behavioural, should be expected among the inhabitants of non-stable environments (Laland et al. 1996; Lefebvre and Giraldeau 1996). Since nomadic-pastoralist systems are usually found in environments that are various and changeable (Dyson-Hudson and Dyson-Hudson 1980, 17), we should expect that the animals participating within these systems will rely increasingly on behavioural patterns that are induced by experience rather than innate. This prediction has indeed been demonstrated to be true for some animals herded by pastoral nomads, including reindeer. In his seminal work on reindeer behaviour, Baskin concluded that:

> The great majority of behavioural complexes exhibited by reindeer is acquired during the individuals' lives. […] Biologically imprinted (in the strict sense) behavioural patterns are few, and those that can be demonstrated do not exist independently. They are always incorporated into learned behavioural models to produce complicated complexes. (Baskin 1968: 6; see also 1970, 2009 for more detailed discussions)

What is more important, however, is that the whole notion of innate as opposed to learned behaviour and, more broadly speaking, of imprinted as opposed to acquired traits, is perceived by contemporary biologists as rather misleading. As one of the leading researchers in animal psychology, Michael Tomassello, vividly explains:

> In phylogeny, Nature selects for ontogenetic pathways that lead to certain results in the sexually mature phenotype. I repeat, Nature selects for ontogenetic pathways that lead to certain phenotypic results. These pathways may rely more or less on the exploitation of exogenous material and information for their realisation and mammals in general […] have evolved many ontogenetic pathways that simply could not develop without such exogenous material or information. […] When developmental biologists look at

> the developing embryo, they have no use for the concept of innateness. This is not because they underestimate the influence of genes – the essential role of the genome is assumed as a matter of course – but rather because the categorical judgment that a characteristic is innate simply does not help in understanding the process or predicting its results. (Tomasello 2001: 49–50)

In other words, it is simply wrong to think that genes program certain behaviour in animals in any simple and straightforward way. The only thing genes can in fact do is send a newly conceived living being along a certain path of development during its life. However, both the process and the results of this development – including the behavioural patterns the animal will end up with – will always depend to a greater or lesser degree on the "exogenous material and information" the animal receives from its environment. This does not mean that genes do not play the role in the formation of this behaviour. Rather, it means that the entire notion of genetic innateness as a rigid and inflexible program that is distinct from learning is misleading.

At least in the case of mammals, the "exogenous information" shaping behaviour can be obtained, broadly speaking, in two distinct ways that were outlined by Hart and Panzer in 1925. The first, called "sub-socially habitual" by Hart and Panzer and "individual" or "asocial" by later ethologists (e.g., Galef 1993; Laland and Hoppitt 2003), is individual learning by trial and error. This kind of learning, in which those behavioural reactions that are rewarded (by obtaining food, escaping various threats, etc.) tend to be repeated again in analogous situations, while those that are punished (by injury, hunger, etc.) tend to be avoided in the future, is well known among all mammals and most other animals. The second way – termed "cultural" by Hart and Panzer and "social" by most modern ethologists (e.g., Heyes and Galef 1996; Heyes 1994; Laland and Hoppitt 2003) – involves the acquisition of information through contact with other individuals of the same species by observing and imitating them. This kind of learning has been demonstrated to be rather common among mammals (although some other animals also exhibit it): nearly all mammals that have observed individuals of their own species being rewarded for certain behaviour (e.g., digging in the ground and finding food) tend to imitate that behaviour themselves. Conversely, an animal that has observed its conspecifics punished for certain behaviour (for example, by dying after eating a certain food or approaching a predator) is unlikely to imitate that behaviour (Heyes 1994).

In line with the intuitions Hart and Panzer had in 1925, "cultural" or "social" learning indeed seems to be effective enough to establish behavioural traditions among animals. These were observed among many animal species whose representatives from different populations exhibit different behaviour (Laland and Hoppitt 2003; Sapolsky 2006). Often these divergent behavioural patterns can be attributed neither to genetic inheritance nor to the differences in environment that can produce different results in individual learning, leaving only differences in social learning to explain them (Laland and Hoppitt 2003; Sapolsky 2006). Some biologists openly advocate the use of the term "cultures" to designate these traditions, while others are reluctant to do so (see Sapolsky 2006 for a review of the debate, which, in our opinion, is purely terminological). A model of the formation of these traditions has been proposed and debated (Boyd and Richerson 1985; Feldman and Laland 1996; Laland 1996; Laland et al. 1996). According to this model, these traditions originate from individual learning by the members of the populations. The behavioural patterns (or the knowledge informing them) discovered by the individual members and found to be rewarding are passed on through social learning and transmitted across the generations from senior individuals to juniors. However, individual trial and error is not inhibited by social learning, and individuals can still look for alternative behaviour. If they manage to discover an alternative that is more effective than the socially learned behaviour, they tend to adopt this in their future behaviour, which can then serve as a model for other members of the population. In this way, the new alternative can be incorporated into social learning and replace the older alternative in the behaviour of the population. By this very process, the behaviour of the members of a population can adapt to a specific environment and, in doing so, diverge from that of other populations.

An obvious conclusion that can be drawn from this discussion is that the behaviour of animals involved in nomadic-pastoralist systems can be and very likely is mutable and adaptable to the environment, a significant part of which, from a biological point of view, is represented by their human masters. Furthermore, it is at least possible that in their reaction to this environment, these animals can form behavioural traditions that are transmittable from one generation to another, and somehow cope with the patterns of actions of their human masters. This possibility is embarrassingly little explored in anthropological studies. The few studies related to reindeer herding that acknowledge this possibility include works of Tim Ingold, Hugh Beach, and Robert Paine. For example, in one of his earlier articles Ingold mentions that reindeer

> also take decisions concerning matters such as pasture access, migratory movements and group status. Undoubtedly, the existence of mankind is recognised in the reindeer decision-making, however rudimentary and short-sighted the latter may be. Men and reindeer thus represent two interacting populations; both form social groups, and are guided in political/economic decision-making, which takes the other into account, by very different sets of goals and values. (Ingold 1974: 521)

Unfortunately, Ingold has not attempted to develop this statement much further on the basis of empirical research. Beach explicitly states that the behavioural patterns of animals can be modified by the actions of herders. These modified behavioural patterns, however, are something that "a herder must impose, usually with considerable coercion, upon his herd" (Beach 1981: 41). As such, they persist only as long as the factors creating them (i.e., the influence of the herder) are present. Once these factors are removed, the animals tend to return to their "natural" behaviour – "a pull like that of a rubber band, as soon as the herder loosens his grip" (Beach 1981: 41).

Therefore, according to Beach, besides adapting their actions to animal behaviour (Beach 1990), herders can also force animals to change their behaviour in a way that better corresponds to the herders' actions (Beach 1981). However, even in this model, only the herders are considered to be active agents. Claims that the relation between animal behaviour and the actions of the herders can be more balanced were made by Robert Paine. He stated that "[a]nimals learn about their herders' order of things, as well as herders about their animals'" (1994: 31), and coined the term "reciprocal learning" to refer to this process, emphasising its balanced character. It should be noted, however, that although both Beach and Paine claimed that modification of reindeer behavioural patterns by the herders was an important factor influencing and maybe even enabling the functioning of reindeer herding systems, neither of them proposed a coherent explanatory model of both the process itself and its relation to herding operations.

In order to address this gap in both anthropological theory and data, we attempted a comparative study of the behaviour of reindeer in the Komi and Nenets herds. This study is particularly appropriate for addressing the observed gap for a number of reasons: first of all, according to the classification of breeds of semi-domesticated reindeer developed by Russian agricultural scientists (Pomishin 1981; Yuzhakov and Mukhachev 2001), reindeer raised by the two groups belong to the same breed (the so-called Samoyedic breed). There can be little doubt that the animals belonging to Komi and Nenets share more genetic characteristics than the animals raised by most other reindeer herding groups (Pomishin 1981; Røed et al. 2008). Secondly, the Bol'shezemel'skaya tundra and the Taz tundra, despite being geographically distant, are still typologically similar as far as their basic ecological characteristics are concerned. Both regions are representative of typical lowland tundra, with essentially the same vegetation set and a comparable dynamic of seasonal changes (Chernov 1980).[77] On the other hand, as was demonstrated in chapter 2, the two groups have rather different pasturing techniques, and their animals are subjected to different herding regimes. These facts allow us to suggest that the differences in reindeer behaviour, if indeed observed, should be related to differences in herding regimes rather than to biological differences between reindeer populations or ecological particularities of the respective territories.

The results of our study are presented and their implications are discussed in the rest of this chapter. Before proceeding to them, however, some remarks on methodology are in order. The evidence on which our argument is built comes mainly from field observations, including participant observation of the reindeer herders' actions toward reindeer and observation of reindeer behaviour. Information about reindeer

77 It should be noted, however, that the climate of the Gydan peninsula (where the Taz tundra is located) is drier than that of the Bol'shezemel'skaya tundra. For this reason, there is a greater difference between the mean annual summer and winter temperatures in the Taz tundra than in the Bol'shezemel'skaya tundra. Still, the mean annual temperature is the same in both places.

behaviour that the herders communicated during ethnographic interviews is also used (we made every attempt to cross-check this information, where possible, with our own observations). We recognise that, from the viewpoint of ethology and animal psychology, observation alone is not considered a sufficiently solid basis on which to make conclusions about animal behaviour: it is usually claimed that such observations should be supported by experiments in controlled environments. However, studies based on such observations are not uncommon in these disciplines, especially in the body of research on animal learning, and arguably have their own advantages over experimental research (for discussions, see Cheney and Seyfarth 1992; deWaal 2006; Mitchell and Hamm 1997). What is more important, observation is arguably the only way to study animal behaviour in the wild and to appreciate the impact of complex environmental factors (Cheney and Seyfarth 1992; Mitchell and Hamm 1997; Mitchell 1997). Furthermore, since our focus is not so much on reindeer behaviour per se, but rather on the interplay between reindeer behaviour and that of the herders, field observations seem to be the necessary initial step of inquiry, providing the basis for building initial hypotheses. This is what we have attempted to do in our study, while recognising that further experimental work is needed to finally prove or disprove our ideas. There is, however, some deviation from the purely observational approach in our comparison of "reaction distances" among reindeer raised by Komi and Nenets herders, as detailed in the next section.

6.2 Differences in behavioural patterns between animals raised by Komi and Nenets herders

Our observations of the behaviour of reindeer raised by the two reindeer herding groups have revealed a set of obvious differences that concern the behaviour of individual animals as well as behavioural patterns followed by animal groupings of various sizes. In the case of individuals, differences in both the type and intensity of reactions to people and dogs were the most pronounced. One way we tested this was by measuring "reaction distance," that is, the distance at which the approach of a human first causes a reindeer at rest to get up and move. In order to perform this comparison, one of us would slowly walk toward a reindeer that was lying on the ground and stop once a visible behavioural reaction was elicited from the reindeer. After that, we would measure the distance in steps to the place the reindeer had been resting before the reaction. All these measurements were performed in summer (between June and August), during periods of rest when the reindeer were ruminating. In the case of Komi reindeer, twenty-four reindeer were tested in this way while they were in the pastures. In the case of Nenets, a total of thirty-three reindeer were tested both in the pastures and in the *tandara* (the area near the camp where the herd is driven after the daily roundup).

The attentiveness of reindeer to the presence of people and dogs in both groups was found to depend greatly on the sex and age of the animal, the presence of offspring, and the distance between an individual animal and others (dispersal of the herd). Nevertheless, it was observed that Komi animals are generally much more attentive to dogs and people than those raised by Nenets. When one approaches to within a distance of five to eight meters from a lone Komi reindeer (in the case of a young reindeer or a female with a calf, eight to ten meters), it causes a reaction. The reaction distance was found to become shorter when reindeer were in a group. However, it was never observed to be less than approximately three meters. On the other hand, when approaching a "packed" herd of Taz Nenets reindeer resting near the camp, one could get as close as one meter or less without provoking any reaction (sometimes, an observable reaction could be achieved only by touching the animal). The reaction distance of a lone reindeer in a pasture was between three and six meters, in other words, approximately half the reaction distance that was observed among the Komi animals. The Komi animals reacted by running toward the nearest group of animals on the pastureland. This normally produced a similar reaction among the members of that group, that is, the whole group would run away to join another group. As a result, in cases where the herd was not too dispersed, "frightening" just one animal was observed to be enough to start a chain reaction resulting in the whole herd gathering together in a single pack. In contrast, the reindeer pastured by the Nenets showed a similar type of reaction only when there was a lone animal or a relatively small group. Medium-sized groups – between twenty and one hundred animals – were just as inclined to run in the direction of the campsite as to run toward other animals. Groups larger than one hundred animals tended to react by running toward the campsite.

A clear difference was also observed in the reaction of large groups (numbering more than one hundred) of running animals to the attempts of herders to change the direction of their movement. In the case of the Komi, frightening the animals at the front of the group (for example, by sending a dog to jostle them or having herders themselves simply stand in front of the animals and block their way) was enough to cause the whole group to turn away and rush in a new direction. On the other hand, among the Nenets herds, a similar technique worked only for smaller groups. When the group of animals was larger (more than one hundred animals), such techniques were more likely to cause the group to split into two or more smaller groups running in different directions.

These differences indicate that the Komi reindeer herds are much more susceptible to direct influence and control by herders than the Nenets herds. This explains why among the Komi a herd can be controlled by just one reindeer herder with a reindeer sledge and dogs, whereas among the Nenets the presence of two herders is required during the short period when their herd is under constant control. During this period we observed that the two Nenets herders divided the work between them in the following way: one herder controlled the forward part of the herd (its "head") and corrected the direction of its movement by stopping and orienting the animals at the front of the pack in a new direction. The second herder controlled the rear part of the herd (its "tail"), forcing it to follow the "head" and preventing it from splitting into smaller groups of reindeer once the "head" of the herd changed its direction of movement. As the reindeer herders put it (in Russian), *Odin krutit golovu, a drugoi derzhit khvost* ("One herder turns the head while the other holds the tail"). Among the Komi, using this terminology, the "tail" of the herd always follows its "head" of its own accord, thereby eliminating the need for a "tail holder."

The reaction of medium-sized and large groups of animals to predators and mosquitoes was also observed to be remarkably different. In the case of the Nenets, groups of reindeer under attack by predators (wolves) tended to stay in a coherent pack and run toward the herders' campsite. The same behaviour is often manifested by small groups of reindeer harassed by mosquitoes, although in this case the animals are just as likely to run against the wind or take refuge on tops of hills. In contrast, among the Komi a group of reindeer under attack by wolves normally dissolves into smaller groups and individual animals that run in different directions. When harassed by mosquitoes, animals tend to move against the wind or toward the tops of hills, or run in circles in one place. We never observed Komi reindeer running toward the herders' campsite under these conditions, nor was such behaviour reported by the herders themselves.

Finally, a very pronounced difference was observed in the behaviour of medium-sized and large (up to 250 animals) groups of reindeer that lost contact with the herd. In the Nenets case such groups, if not found by the herders during the course of the daily roundups, would return to the campsite of their own accord after some time. The herders reported that these reindeer tended to return to the same campsite that they had been driven to by the herders during the previous roundup. When the herd is rounded up and moved to a new campsite, those animals that are missing will most likely return to the previous campsite. It was also reported that the certainty of this behaviour depended on the number of times the animals had been driven to a particular campsite. For example, if the animals were "lost" immediately after moving to a new site, or if they had been rounded up and driven to this place only once or twice, the chances were that they would return to the previous rather than the current campsite. The validity of the herders' reports was confirmed by observation. Furthermore, it was observed that Nenets herders relied heavily on this reindeer behaviour in their herding practice. For example, in spring and early autumn, when the pasture conditions were good and reindeer did not wander far away from the camp for long periods of time, the Nenets herders tended stay in one place for longer periods and, therefore, were more accepting of large herd dispersals before moving on to a new place than they were in summer and winter. Their explanation was that reindeer had "learned" to return to this particular place for long enough, and the danger of losing them was minimal even if the herders were unable to find and round up all the animals every day. The animals that the duty herders failed to find were likely to return to the camp by themselves eventually. In one particular case in June 2006 we observed that after a twenty-two-day period of staying in one particular place, the herd became so dispersed that the duty herders only managed to round up approximately one-third of the animals every day. The camp was not moved, however, and the herders expected all the reindeer to return to the campsite of their own accord when mosquitoes started to harass them at the beginning of the summer. This expectation turned out to be justified.

No similar behaviour was observed among the Komi herds. Reindeer that had lost contact with their herds did not tend to move to any particular place, but rather to roam more or less freely across the tundra. Although the Komi herders did report cases in which individual reindeer had returned to the herd of their own accord after some period of time (sometimes quite a long period), they clearly stated that such cases are so exceptional that these animals are given the nickname *asloktys'* ("self-returner"), which remains with them until their death. In the three brigades of Komi reindeer herders that we visited in the course of our fieldwork, there were only three animals with this nickname. Furthermore, the Komi herders did not report any instances of large groups of animals returning to their herds of their own accord and, when asked about the possibility of such an event, found it improbable or even impossible.[78] The behavioural differences detected between the reindeer pastured by the two groups are summarised in table 3.

Behavioural characteristics of reindeer	Komi of Bol'shezemel'skaya tundra	Nenets of Taz tundra
Distance at which individual reindeer react to people and dogs	5–10 meters	< 3 meters near the camp; 3–6 meters on pastureland
Reaction of small groups of reindeer (20–100 animals) to people and dogs	Run toward the nearest group of grazing animals	Run toward the nearest group of grazing animals **or** toward the campsite

78 As one of the older informants explained, an individual reindeer that has become detached from its herd feels anxious and roams across the tundra in search of a herd of reindeer it can join. It might, therefore, find its own herd after some time. Groups of animals, on the other hand, do not actively search for a herd or do so less intensively, making the probability of such groups returning to their herds much smaller.

Reaction of large groups of reindeer (more than 100 animals) to people and dogs	Run away as a coherent group	Split into smaller groups
Reaction of medium-sized and large groups of animals to an attack by predators	Dispersal	Run toward the herders' campsite as a coherent pack
Reaction of medium-sized and large groups of animals to mosquitoes	Run uphill, against the wind, or in circles	Run uphill, against the wind, or toward the herders' camp
Behaviour of individual animals or small groups that have lost contact with the herd	Roam more or less unsystematically in search of the herd	Return to the campsite where the animals had been driven most recently

Table 3. A short summary of the observed differences in the behaviour of reindeer raised by Komi and Nenets herders.

These observations confirm that there are indeed differences in behavioural patterns between the reindeer managed by these two reindeer herding groups. Taking into account the essential similarities of the genetic structure of the two reindeer populations and the ecological characteristics of their habitats, these differences are most likely related to the different herding technologies to which the animals are subjected by their human masters. Indeed, reindeer behaviour is likely to be in many respects the product of a specific reindeer herding technology. As demonstrated in the first section of this chapter, this finding is exactly what could have been predicted on the basis of the existing ethological literature. Much less predictable is the fact – also demonstrated by the data presented here – that herding technologies themselves seem to be essentially dependent on reindeer behavioural patterns. As the examples of the Nenets waiting for their reindeer to return to the campsite and of lone Komi herders controlling their herds (with a reindeer sledge and dogs) clearly indicate, the herders seem to act on the assumption that these patterns are already present. Furthermore, their actions seem to be effective precisely *because* these patterns are present. This becomes especially visible in those cases when the herders try to apply their herding technology to animals that do not demonstrate the required behaviour. For example, during fieldwork, the older Komi herders distinctly recalled the problems they had with reindeer that were purchased for them from the Yamal Peninsula at the end of the 1950s, when they experienced extensive losses of livestock due to foot-and-mouth disease. According to their descriptions, the new animals were completely unmanageable, liable to escape and destabilise the entire herd by leading groups of "old" reindeer away with them. As one of the older herders explained, "One could not know what to expect of them." As a result of this behaviour, most of these animals were lost in the first few years after their introduction. The Nenets herders told similar stories about *hargin*[79] reindeer from Chukotka. Several attempts to introduce small numbers of these reindeer into their herds were made at the end of the 1970s and the beginning of the 1980s in order to "improve" the physical characteristics of the local reindeer (i.e., increase their size and weight). In all of these cases, the imported reindeer were characterised as "wild." Based on the previous discussion, it can be supposed that the newly introduced animals, which were raised in and behaviourally adapted to a different reindeer herding system, simply lacked the behavioural patterns required for the normal functioning of the new reindeer herding system. They got lost before they had a chance to adapt their behaviour to the new system.

6.3 Dynamic mutual adaptation

The evidence presented in the previous section confirms that reindeer behaviour is flexible and can change in response to environmental factors. Reindeer are not different in this respect from other mammals. Probably the most important environmental factor in the case of all semi-domesticated animals is the actions of their human masters. The range of possible behavioural patterns of any animal is restricted. The studies mentioned earlier in chapter 2, the first section of this chapter, and the data presented in the previous section suggest that the animals cannot simply be forced to behave in the way a herder finds appropriate. However, the converse statement implied by some of the studies reviewed, i.e., that herders constantly adapt their behaviour to that of their animals, also seems incorrect in the light of the material presented.

Our observations suggest that the relation between the patterns of animal behaviour and the patterns of the herders' actions should be viewed as a dynamic mutual adaptation. In general terms, this means that by modifying their behaviour to adapt to herders' actions, the reindeer create the necessity for the herders to adapt their actions to that modified behaviour, which, in turn, produces a new modification in animal behaviour. Since there is always more than one way to adapt successfully to a new behavioural pattern on the part of the herders and, probably, that of the animals, this loop can and does produce differences in both animal behaviour and herding technology between different groups of herders.

This general model is actually a more detailed rendering of the interactionalist concept of reciprocal learning suggested by Paine. However, it also takes Paine's

79 The term *hargin* (also: *khargin*) refers to the breed of reindeer developed and pastured by Chukchi and Koriak reindeer herders of eastern Siberia. Compared to other breeds of reindeer, *hargins* have a significantly larger body mass and, therefore, greater meat productivity (exceeded only by the woodland reindeer raised by Tofa and Tozhu herders in southern Siberia), and they are significantly less dependent on lichen during the winter.

argument (1994: 31) that "animals learn about their herders' order of things, as well as herders about their animals'" a bit further. It emphasises that the "herders' order of things," which the animals learn, is itself the product of the herders' learning of the animals' order of things. Indeed, the aspects of reindeer behaviour observed among the Komi and Nenets can be easily accounted for by applying this general approach. Thus, it could be suggested that the return to a campsite, demonstrated by the Nenets reindeer, must have been the product of their tendency to rejoin the herd, which has been modified by the animals' experience with a particular herding technology that includes the practice of regularly rounding up the herd and driving it to the campsite. This practice allows the reindeer to learn the focal point where other animals are most likely to be met and where safety from predators and other disturbing factors can be found. On the other hand, it could be suggested that Komi reindeer, which are constantly "packed together" by the duty herders, have learned that being on the inside of a large herd, along with the maximum possible number of other animals, is the easiest way to avoid being further disturbed by the herder and his dog. This explains the tendency of animals in Komi herds to readily gather together and to react to the actions of herders as a coherent group. In both cases, however, these learned patterns of behaviour give the herders the possibility of changing their patterns of actions to make them more efficient. Thus, the Nenets rely on the learned behavioural pattern of their reindeer and are spared the need to look for and round up the whole herd in certain periods. It is even feasible that the very possibility of maintaining only intermittent control over the herd without much loss of livestock is created by this learned pattern. On the other hand, the learned behaviour of the Komi reindeer allows the herders to economise significantly on the work required to control the herd.

On a more general level our model predicts that any particular herding technology is likely to produce a specific "tradition" of animal behaviour that then becomes essential to the functioning of that technology. In fact, our material does not allow us to say with confidence whether the modified behavioural patterns spread among the animals solely through asocial learning (i.e., each individual animal's independent discovery by trial and error of a similar response to the herders' pattern of actions), or if social learning (i.e., observation and imitation of other animals' behaviour) also plays a role. In any case, the behaviour of experienced animals does of necessity play a central role in younger animals' learning processes, even if only by supporting the possibility and effectiveness of the herders' actions that induce particular behavioural patterns. Indeed, even if each young Nenets reindeer independently discovers the usefulness of returning to the campsite, it does so only because the Nenets technology induces the regular concentration of reindeer near the campsite. On the other hand, as has been demonstrated, the Nenets technology can induce such a concentration of animals because the corresponding behavioural patterns exist among the experienced reindeer. In this way the behavioural traditions constantly reproduce themselves by social means either directly (by younger animals' social learning from more experienced animals) or indirectly (through the younger animals' independent learning from the situation created by experienced animals). In this sense at least we can justifiably speak of reindeer "cultures" always accompanying the cultures of the herders.

In the cases of the Komi and the Nenets, it appears that dynamic mutual adaptation has led to the increased efficiency and stability of reindeer pastoralism. However, the animals and herders can also fail to successfully adapt their behaviour to one another, in which case a vicious circle of dynamic mutual *maladaptation* can lead to the destabilisation of the pastoralist system. A process of this kind occurred among the Skolt Saami of northern Finland after the introduction of snowmobiles, as described by Tim Ingold:

> The use of snowmobiles in herding further altered the man-deer relationship. When deer were driven by skimen with dogs, [...] they moved largely "under their own steam" at a natural pace. On the other hand, the loud noise and the speed of the snowmobile terrify deer. [...] Deer have learned to avoid "snowmobile men" at all costs and to "hide" in the most inaccessible parts of the forest. Thus, the former "transactional" man–deer relationship, first breaking down when the pastoralist could no longer fulfil his side of the bargain, has become one of pursuer and pursued. Loss of control over deer leads to a drop in the proportion that are actually found each winter. (Ingold 1974: 529)

A more recent example of a similar process can be found on the Kola Peninsula, just to the west of the regions described in this paper. There, according to Vladislava Vladimirova (2006), the herders' gradual abandonment of constant observation of their herds during the summer for the last several decades led to a greater dispersal of reindeer over the tundra in the summer and early autumn. This greatly increased the time and effort required to recover the herds in the late autumn and early winter period. As a result, the period during which the animals were not under the control of the herders grew to include a part of the winter. This in turn caused a further increase in the dispersal of the reindeer, and the herders had to search for, round up, and drive for long distances (and, later, corral, count, and deliver injections to) the female animals that were in advanced stages of pregnancy and which, in order to protect their fetuses, had learned to avoid the operations by escaping into distant parts of the tundra. The result was a further delay in taking control of the herds. Currently, the herders manage to collect their animals only by the end of March or the beginning of April, which is only one month before the calving period. The animals then tend to disperse immediately after they are released from the counting/injection corral and escape deep into the tundra. As could be expected, this has led to the extensification of local reindeer herding and a sharp increase in livestock losses.

We believe that our explanation of the Kola and Skolt Saami situations as examples of dynamic mutual maladaptation has certain advantages over the model Hugh Beach (1981) developed to explain a similar process in Scandinavia. Indeed, the central assumption of Beach's model is that reindeer have certain "natural" behaviours, that is, a set of naturally (genetically?) given behavioural patterns to which they tend to return, all other factors being equal and, in particular, if the impact of humans is removed. Therefore, "there is always a constant leaning towards extensivity" (i.e., a loosening of the bond between humans and animals – Beach 1981: 41) in a reindeer herding system, and this explains the transition to extensive herding. However, as the discussion in the first section of this chapter suggests, the existence of "natural" reindeer behaviour in the sense Beach seems to have in mind is rather doubtful since, among other things, other factors simply cannot be equal. Various and varying environmental factors, including the impact of human beings (which can never be completely removed), will always affect reindeer behaviour. If there is indeed anything "natural" for reindeer, it is their ability to adapt their behaviour to changes in environmental factors. No "programmed" direction of such change is likely to exist. The new behavioural patterns of the reindeer from the Kola Peninsula or northern Finland are not more "natural." Rather, they are simply an adaptation to a new mode of herders' actions that happens to be disadvantageous for the herders and to which they have failed to adapt.

6.4 Implications of the dynamic mutual adaptation model for the process of deciphering and predicting reindeer behaviour

The material presented in this chapter and the model proposed to explain it allow us to suggest some modifications to the theoretical treatment of the concept of domestication with regard to reindeer. It is well known that reindeer are not domesticated animals from a biological or physiological point of view. In other words, no significant biological differences between domesticated (controlled in any sense by humans) and wild (uncontrolled by humans) reindeer can be demonstrated. This fact has recently caused the introduction of the term "semi-domesticated" when referring to controlled reindeer. No commonly accepted definition of this term exists at this time. However, Beach and Stammler (2006) have recently claimed that it actually refers to behavioural differences between controlled and uncontrolled reindeer. Our material supports this claim, but with an important correction. Indeed, our model predicts that reindeer behaviour becomes modified once the animals become part of a pastoralist system. However, it also predicts that it becomes modified in different ways depending on the herding technology. Also, as has been mentioned, the existence of any single set of behavioural patterns common to all non-controlled populations of reindeer is doubtful from an ecological point of view. Our model furthermore predicts that behavioural patterns present in all populations of controlled animals and absent in any population of uncontrolled ones are highly unlikely to exist. Hence, we believe that "semi-domestication" (in contrast to "domestication" proper) can only be relative to a pastoralist system. In other words, the term "semi-domesticated reindeer" can be best understood as referring to animals whose behaviour patterns are adapted to those of the herders' actions within a particular reindeer herding technology. By extension, this would mean that the same population of reindeer can be "semi-domesticated" in one system but remain "wild" in another.

This outlook, in our view, corresponds closely to the way the terms "domestic" and "wild" are used by reindeer herders themselves. As already mentioned, the herders commonly refer to animals newly introduced from other herding systems as "wild." This means, it would seem, that their behaviour cannot be dealt with in the framework of the herders' particular pastoralist system, and that it does not make sense to the herders. In other words, for a reindeer herder to say, "This reindeer is wild," is likely to mean simply, "This reindeer does not behave in the way I expect it to," or (which is probably the same), "I cannot make sense of (i.e., decipher and predict) its behaviour." It is important to bear in mind that this does not necessarily mean that a particular reindeer indeed comes from a population uncontrolled by humans, as the example of Komi herders believing that Yamal reindeer were "wild" clearly suggests. This situation can occur just as easily if the reindeer in question has been raised in and behaviourally adapted to a reindeer herding system that is different from the one of the herder.

This suggests, in turn, that the dynamic mutual adaptation of reindeer and herders to each other is likely to be the necessary precondition of the herders' ability to decipher and predict reindeer behaviour. In other words, the herders are able to predict animal behaviour by virtue of knowing the specific "culture" of their reindeer – the socially transmitted models of reindeer behaviour – that has emerged in response to (and therefore has been shaped by) the "reindeer-herding culture" – the patterns of dealing with the reindeer – of the herders themselves. It can be suggested, furthermore, that this "culture" of reindeer makes sense to the herders exactly because it represents, in a very important sense, a product of their own actions. This means that the ability of herders to understand and predict reindeer behaviour emerges from the fact that this behaviour, thanks to the dynamic mutual adaptation process, is itself organised in such a way as to make sense in the framework of the herders' action.

Of course, this conclusion does not by itself say anything about the cognitive mechanisms that reindeer herders can employ in order to explain and predict the behaviour of their animals. It does, however, put the question of the possible design of these mechanisms in a new light. First of all, it suggests that these mechanisms can and probably do rely on objective relations between reindeer behaviour and the actions of reindeer herders. Secondly, our conclusion allows us to approach these mechanisms from a different perspective. Indeed, quite a common position in anthropology (discussed in greater detail in the next chapter) is that the ability of aboriginal people to understand, predict, and exploit animal behaviour is based on their so-called traditional ecological knowledge (TEK). This knowledge is often portrayed as consisting of general statements describing some regularities (not dissimilar to scientific laws) of animal behaviour and the environment in general. Many researchers would argue that these "laws," although probably more limited in scope than the laws of modern science, are products of observation and experimentation performed by aboriginals in the course of "countless generations" (see Huntington 2000 for review and criticism). The aboriginals are therefore viewed as acting very much as modern scientists in building up their knowledge about the environment and its components: they start by collecting empirical observations, then build up hypotheses that explain causal relations between the observed phenomena, and, finally, test these hypotheses in the "experimental settings" related to their everyday activities. The material presented in this chapter suggests that, at least as far as reindeer herders making sense of reindeer behaviour is concerned, there can be a different way to achieve and organise the necessary knowledge. Instead of acting as natural scientists, the reindeer herders can act more like ethnographers and try to make sense of the "culture" of their animals through a kind of "participant observation" that leads to a "hermeneutic circle." In other words, the herders can learn to make sense of their animals by observing and, to a degree, shaping the animals' reactions to their own actions, trying all the while to decipher the animals' point of view in the process. The degree to which this is possible is the subject of the next chapter.

7 THE THEORY OF REINDEER MIND: A COGNITIVE EXPLANATION OF HERDERS' ABILITY TO DECIPHER AND PREDICT REINDEER BEHAVIOUR

7.1 Traditional knowledge and traditional skills: Anthropological explanations of deciphering and predicting animal behaviour by aboriginal people

It would be appropriate to start this chapter with a small anecdote that vividly illustrates how reindeer herders routinely decipher and predict reindeer behaviour and the importance of this ability in the everyday lives of the herders. On a rather warm August evening in 2006, one of us observed how the Taz Nenets reindeer herders "pushed away" the reindeer herd from their campsite after a daily roundup (see chapter 2). Usually after each roundup, two to four male herders – those who would perform the next roundup – lasso three transport reindeer each and harness them to their sledges in order to have them ready to for the next roundup. This time, however, was different. Although the herders who had just performed the roundup released their transport reindeer so they could rest, no fresh animals were caught and harnessed. This meant that, as the herd was leaving the campsite to disperse in the tundra, the herders were left without any transport animals that could be used to search for the animals the next morning. Knowing that it was next to impossible to round up a dispersed reindeer herd without reindeer sledges, the researcher turned for explanation to his key informant, an experienced herder named Mikhail Salinder, who observed the situation with a look of disapproval on his face. "They believe that it will be as hot during the night and tomorrow morning as it is now," he explained. "The mosquitoes will harass the herd. The reindeer, wanting to get a break from them, will come to the camp by themselves … because they will reason that the smoke from our fires will drive the mosquitoes away." Noticing a surprised look on the face of the researcher, he added, "The reindeer will know that our campsite is the nearest place without mosquitoes."

> "But this is dangerous! What will we do if the reindeer do not come?"
> "Indeed, shamans they are!" agreed Mikhail with sarcasm. "How can they know for sure what the weather will be like tomorrow?!"

Of course, anyone would probably agree that the old herder had reason to be worried. One can never be completely sure that the weather will not change in the next twelve hours, even if, as was the case, there was neither wind nor clouds nor any other signs of a sudden change in weather. Still, most people from modern industrialised societies, including ours, would probably find this particular assumption – that weather is not going to change – the safest in the chain of conclusions the old herder described. What would appear much more incomprehensible to an "industrialised person" is how such a potentially dangerous decision to skip lassoing transport reindeer could be based on assumptions about what the reindeer would want, know, and, most surprisingly, how they would reason! Still, it seemed like these were the assumptions none of the herders, including Mikhail, felt any need to take issue with. In any case, all the predictions did work out perfectly: the weather did not change and the animals did indeed come to the campsite by themselves the following day.

Stories like this can be found in abundance in the anthropological literature. In different parts of the world, hunters, herders, and occasionally agriculturalists have amased anthropologists both by their ability to explain and predict animal behaviour, and by their frequent reference to the animals' desires, beliefs, and reasoning while doing so. Interestingly, these two cognitive aspects have most often been treated separately, each being viewed as demanding an explanation in its own right. Thus, the ability to decipher and predict animal behaviour is frequently taken as an example of the detailed knowledge that aboriginals possess about their environment (Berkes et al. 2000; Bock 2005; Liebenberg 2001; McDonald 2007, to name but a few recent comparative and review works). Reference to the mental states of the animals, on the other hand, has frequently led to extended discussions of "animism" and "anthropomorphism" as part of both "traditional" and "modern" worldviews (see Caporael 1986; Ingold 2000: 111–131 for a recent treatment), and has been believed to be essentially unrelated to the ability to decipher and predict animal behaviour per se.[80] One of the arguments we wish to make in this chapter is that this separation is deeply flawed because anthropomorphism is an integral part of the ability to decipher and predict animal behaviour, and is in fact a highly efficient tool for doing so. However, before making this and several other arguments regarding the cognitive mechanisms behind the herders' ability to decipher and predict reindeer behaviour, it is worth reviewing briefly how this ability has been approached and explained in cognitive anthropology to date.

As early as 1939 Bronslaw Malinowski asserted that:

> To every type of standardised technique there corresponds a system of knowledge embodied in principles, which can be imparted to those who learn, and which help to co-operate those who are already trained. Principles of human knowledge based on true experience and on logical reasoning, and embodied in verbal statements, exist even among the lowest primitives. (Malinowski 2006 [1939]: 97)

The main idea behind this assertion is that knowledge takes the form of a system of "principles embedded in statements" (i.e., "propositions," as more recent researchers prefer to say), which refer to the way two or more perceived components of the

80 A recent monograph by Rane Willerslev (2007) represents an important exception here.

environment are, supposedly, related in the objective world. In other words, human knowledge is a system of statements of the following general form:

(1) [Under condition C] A is related to B in the way γ.[81]

A, B, and C stand for perceived components of the environment (the condition C can be absent if the relation between A and B is believed to exist under any condition), while γ specifies the exact mode of the relation – it can stand for causality (A causes B), mode of process (A behaves in the way B), transformation (A takes the form B), etc. The basic assumptions of this model of knowledge have survived to the present day not only in cognitive anthropology, where the system of propositions of the general form (1) is considered to be one essential type of cultural models, but also in sociocultural anthropology in general. Thus, as recently as 2002 Frederick Barth argued that the central component of traditional knowledge is "a corpus of substantive assertions … about aspects of the world" (Barth 2002: 2). His examples of these assertions suggest that he actually has in mind propositions of the general form (1).

Of course, the model itself, as it is applied in cognitive anthropology, has changed since the time of Malinowski. The most important transformation occurred with the realisation that the items related by propositions are not in fact perceived components of the environment, but concepts and categories into which people group the infinite variety of phenomena of the experienced world (Atran 1990; Ellen 2006; Tyler 1969). These concepts and categories are at least partly cultural products themselves (Atran 1990; Ellen 2006) and, therefore, they form an inseparable part of cultural models, including culturally acquired declarative knowledge. Thus, the model of knowledge came to consist of two large parts: 1) the body of concepts or categories into which the flow of experience is sorted; and 2) the body of propositions that relate these concepts to one another and provide the basis for practical and theoretical inferences about the world. A number of other types of cultural models, such as "scripts" (Schank and Abelson 1977), "rules of thumb" and decision rules (Cronk 1991; Smith 2000), emotional patterns (Geertz 1983; Quinn and Holland 1987), and others (see Antweiler 1998, 2004 for a more comprehensive review), have also been proposed as a part of the knowledge system.

From the viewpoint of this general model, the ability to predict animal behaviour demonstrated by the Nenets herders in the anecdote above could be understood as follows: First, the herders observed the environment and classified the existing weather conditions as falling under the category "warm day." They then take an informed guess that the same weather conditions will continue until the following day. The remainder of the process could consist of the following chain of evoked propositions and inferences achieved on this basis (table 1).

81 The definition of the general form in this way broadly follows the discussion in Smith 2000: 30.

Proposition evoked	Inference made
At the end of August (C), if the day is warm (A), there is always an abundance of mosquitoes (B).	There will be abundance of mosquitoes tomorrow morning (deduction).
[Irrespective of all possible conditions (C)], when mosquitoes are abundant (A), reindeer always move to a place where the number of mosquitoes is limited (B).	Tomorrow morning the reindeer will move to a place where the number of mosquitoes is limited (deduction).
[Irrespective of all possible conditions (C)], the number of mosquitoes (A) is limited in our campsite (B).	Tomorrow morning, reindeer are likely to come to our campsite (induction).

Table 1. Predicting the return of reindeer to the campsite from the viewpoint of the body-of-propositions model.

As one can see from table 1 (and as was also mentioned toward the end of the previous chapter), making predictions in this model is strikingly similar to inferring particular predictions from a general theory in science (see Popper 1992): aboriginals are presumed to base their predictions on more general statements describing some regularities (not dissimilar to scientific laws) of animal behaviour and environment in general. Their ability to solve complex cognitive tasks in this model depends, therefore, on the richness and detail of the body of propositions describing these regularities. In particular, the ability to predict animal behaviour presupposes the knowledge of an abundance of detailed "rules" or "laws" of this behaviour under different conditions.

In the previous chapter we already started to argue that this is not an adequate picture of the process of reindeer herders predicting and deciphering animal behaviour because the "laws" and "rules" of reindeer behaviour are likely to be products of a dynamic interaction between herders and animals, rather than natural regularities the herders could infer, for example, through a kind of statistical learning. To this argument we can now add that the idea of predictions as inferences from general rules in this particular case is flawed also from a purely theoretical point of view. Indeed, as the extensive criticism of stimulus-response behaviourist ethology tells us (e.g., Cheney and Seyfarth 1992; deWaal 2006, 59–67; Mitchell et al. 1997), it seems rather unlikely that the details of behaviour of mammals (reindeer included), or most other animals for that matter, can be described by a manageable list of propositional "laws." Such laws can only describe at best some very general behavioural tendencies, such as that certain birds fly south in the autumn and north in the spring. It is hardly surprising, therefore, that despite both the massive conceptual taxonomies (e.g., Berlin 1992; Hunn 1977) and the rather extended lists of propositions that have been collected in different aboriginal contexts, actual instances of explaining and predicting animal behaviour from them has rarely been observed (Ingold 2003; Lave 1988, 1996; Nygren 1999). Furthermore, the great majority of these propositions seem to be informants' *post-hoc* explanations of how particular inferences were made, rather than the actual basis of these inferences (Ingold 2000) – a fact that would seem to explain the numerous contradictions often observed between propositions supplied by the same informant.

The recognition of these facts has led to growing dissatisfaction among anthropologists with the body-of-propositions model of knowledge. It has been argued that the inferences of aboriginals (as well as other people in everyday situations) cannot be based exclusively (if at all) on a body of explicitly or implicitly held general-order propositions (Lave 1988, 1996; Nygren 1999). In fact, the very existence of such inferences has increasingly been questioned (Bourdieu 1977; Ingold 2000; Lave 1988). It has been argued, rather, that knowledge is essentially situated; that is, it is inseparable from the situation in which it is applied (Lave 1988) and the action in which it is involved (Ingold 2000; Lave 1988). This knowledge, therefore, is implicit and can be explicitly reflected upon by informants only to a greater or lesser (but never absolute) degree of accuracy (Bourdieu 1977; Lave 1988; Nygren 1999). In other words, traditional knowledge is largely processual rather than declarative, and can be better understood using the notion of "skill."

These arguments appear to be both substantial and invincible. They clearly explain why previous attempts to understand the abilities of indigenous peoples have failed. What they do not tell us, however, is how (if at all) we can endeavour to understand them. Indeed, it seems that, for many of the authors, to claim that natives (as well as other people, for that matter) are doing something using their situated skills is already an explanation of their actions. However, this claim alone surely cannot be taken as a meaningful solution to the problem. Indeed, if we are to continue our project to understand how and why people live the lives they do, we should at least attempt to understand the mechanisms on which these skills are based.

7.2 Interpreting and predicting animal behaviour through the attribution of mental states

If one cannot predict animal behaviour on the basis of unambiguous rules and laws, then how can such a prediction be possible at all? Indeed, it could well be a product of an implicit and situated skill. However, this skill should necessarily explore and exploit some information contained in the human brain and/or distributed in the environment. Reflecting on what this information might be is a good way to start to understand the mechanism behind the ability.

We would argue that making detailed predictions about animal behaviour is possible because this behaviour, despite defying description in the form of a list of rules and laws, is *not random*. The main reason why it is not random is that it is *purposeful*, that is, it aims at satisfying the needs of the animal and it is guided by the relevant *information* the animal has been able to obtain about its environment. This statement may appear rather obvious, but it suggests a good way to achieve the prediction under question: one can predict animal behaviour by treating the animal as what Daniel Dennett has called an "intentional system," and by adopting an "intentional stance" toward it (Dennett 1981: 1987). In this case,

> One predicts behaviour [...] by ascribing to the system the possession of certain information and supposing it to be directed by certain goals, and then by working out the most reasonable or appropriate action on the basis of these ascriptions and suppositions. It is a small step to calling the information possessed [the system's] beliefs, its goals and subgoals its desires. (Dennett 1981: 6)

In other words, one can achieve the prediction by attributing to the animal certain mental states – desires, beliefs, and, we would add, emotions – that the animal has at the moment or is likely to have in the future, and by inferring an optimal pattern of intentional action from these mental states.

We would like to stress that this way of achieving predictions of animal behaviour is by no means purely theoretical. Quite the contrary, it seems as if this way of making sense of and predicting animal behaviour (or, perhaps, even any kind of behaviour) is most natural for humans; it is their "cognitive default" (Caporael and Heyes 1997; Caporael 1986; Povinelli 1997). It has even been claimed that humans, thanks to particularitiesof their cognition, simply cannot take any other approach toward animal behaviour[82] than to make sense of it through the attribution of mental states (Asquith 1984). In any case, from a phenomenological point of view, the "intentional stance" model seems to be a much more plausible account of predicting animal behaviour than the body-of-propositions model described in the previous section. Indeed, consider a simple case of predicting the behaviour of a rat, placed in a simple laboratory environment, after several hours of water deprivation.[83] One need not be a specialist in rat ethology to predict that this rat will press the bar of its drinking vessel at an increased rate or that it will spend more time drinking. In fact, any person could make a number of other correct predictions about the laboratory rat. Now, would it be in any sense correct to say that we achieve these predictions by consulting a body of propositions that we have and picking out a proposition such as, "In laboratory conditions and after several hours of water deprivation (C), the behaviour of laboratory rat (A) takes the form of pressing the drinking-vessel bar at an increased rate (B)"?

82 at least as far as more or less anthropomorphic animals are concerned.

83 The idea of this example was inspired by the discussion in Whiten 1996.

In our opinion, it is highly unlikely. Rather, what we really do is figure out that after several hours of water deprivation the rat is very likely to be *thirsty*. In other words, we *attribute a mental state* to the rat to *interpret* its behaviour.[84] All the predictions we then make are achieved based on this attribution, as well as on our belief that the rat is likely to act purposefully on it.

On the other hand, although deciphering and predicting animal behaviour through the attribution of mental states is likely to be very natural for humans, it has been repeatedly claimed by animal ethologists to be deeply flawed (see, for example, Kennedy 1992 for a recent extensive criticism). The main reason why it is believed to be flawed is that it is thoroughly anthropomorphic; as such, it is based on the unjustified attribution of human qualities to non-human animals. Indeed, although animals do act purposefully, it is rather doubtful that they can have mental states analogous to or resembling our own. In other words, it is hardly possible for a rat (or a reindeer) to have desires (for example, to be thirsty), experience emotions (for example, anger), or entertain beliefs in our sense of these words. Therefore, attributing such mental states to the animals is in and of itself an unwarranted anthropomorphism, which, according to the critics of anthropomorphism, is very likely to lead to serious miscalculations. The situation gets even worse when we not only attribute mental states to the animals, but also apply our views of the dynamics of mental states in order to infer how these states are likely to change. An example would be attributing a certain belief to an animal and then predicting that it would have another belief that we believe can be inferred from the first (e.g., "Since reindeer believe that P, then they would reason that Q"), which is frequently used in predicting behaviour through the attribution of mental states. Another example is the reasoning (involved in the thought experiment that we have just described) that a rat that is deprived of water for several hours is likely to be thirsty. Indeed, even if a rat can have a mental state bearing some resemblance to what we would call "feeling thirsty," the only reason we have to think that the animal will have this state after several hours of water deprivation is that we humans would have it in this situation. This is, of course, another case of unwarranted anthropomorphism. Therefore, the animal ethologists would argue, attributing mental states

84 One might wonder if concluding that the rat is thirsty indeed amounts to attributing a mental rather than a physiological state to it. We argue that it does for the following reason: our predictions about the behaviour of the rat are based not only on our conclusion that the rat is physiologically dehydrated but, more importantly, on our suggestion that the rat feels (subjectively experiences) that physiological state as a mental state. Indeed, if we were told that the rat was unable to experience this physiological state as a mental state – that it was unable to feel thirsty due to a certain experimentally induced brain lesion, for example – we could still say that the rat was thirsty but our predictions about its behaviour would likely be very different. Similarly, although we can say "he is thirsty" about a dehydrated person lying in a coma (and unable to feel anything), we would not use this attribution to explain the behaviour of this person. This proves that our prediction is based on attributing to the rat a mental state (i.e., "feeling thirsty") rather than a physiological state (i.e., "being dehydrated").

to animals, however natural it may be for us to do so, cannot be an effective way to make sense of or predict their behaviour.

In order to respond to these doubts it is worth mentioning, first of all, that any doubts that animals are capable of having mental states in "our sense" can be justified only if we have a clear idea what mental states in "our sense" actually are. Unfortunately, this clear idea is exactly something we manifestly lack. The nature of our mental states has long been debated by philosophers of mind, and there is no indication that this debate will be resolved anytime in the near future. It is beyond the scope of this book to review here all the numerous positions taken or expressed in the course of this debate. It is worth pointing out, however, that one (and by no means the most marginal) of these positions, usually called "instrumentalist" and embraced by, among others, Daniel Dennett (1981, 1995) and Dale Jamieson (2009), is that "for a creature to have mental states is simply for its behaviours to be richly and voluminously predicted by the principles of our folk psychology" (Lutz 2009: 5). If we take this as "our sense" of having mental states, then it can be concluded that not only animals, but even certain kinds of machines – such as a computer or a thermostat (see Dennett 1995: 114) – can have mental states similar to ours.

However, even if animals indeed do not have any mental states "in our sense," it does not follow that their behaviour cannot be correctly predicted by attributing such states (and certain assumptions about their dynamics) to them. As Gallup and colleagues note:

> [I]respective of whether the dog really loves his master or not, I might be able to develop a predictive model of the dog's reactions to his owner based on the use of a "love" metaphor. In the latter case, it is important to emphasise that in order for anthropomorphism to have predictive value it does not have to be a valid explanation for why the animal is doing what it is doing. (Gallup et al. 1997: 80)

Indeed, in the very worst case the attribution of a mental state on the basis of observable behaviour can work as a generalisation from empirical data to an abstract theoretical concept, while the prediction of the behaviour would be not dissimilar to making an empirical prediction on the basis of a theoretical concept (see Whiten 1996 for a similar point). To the degree that the relations between certain empirically observable facts (e.g., behaviour) and abstract theoretical concepts (e.g., mental states) are correctly surmised, it is possible to predict the former using the latter irrespective of how the latter are related to any other facts such as the phenomenological or neural states of the animal. Indeed, consider a physicist predicting the trajectory of a satellite's movement on the basis of the satellite's mass. It is rather certain that the satellite cannot experience phenomenologically its mass in the same way that we can experience ours. One could justifiably argue, therefore, that the satellite cannot have mass "in our sense." Still, the physicist certainly can achieve a reliable prediction of the satellite's trajectory because, thanks to the works of Newton, he has got the important relations – those between the mass of the satellite and its movement – right, while the other relations – those between the mass and the phenomenological experience – can be safely ignored in this case. Of course, both the possibility and correctness of prediction in this case would depend on the quality of the body of theory that relates the observable phenomena to the abstract concepts. In certain cases, one would need quite large and complicated theories. Still, such theories almost certainly would be more effective and manageable than the lists of propositions relating one set of observable phenomena to another as postulated by the body-of-proposition model (see Whiten 1996 for arguments).

We would argue, however, that at least for some animals the mental states attributed can be more than just abstract theoretical concepts. Indeed, it seems obvious that our ability to experience emotions, have desires, and entertain beliefs is a product of natural selection that shaped our nervous system in a certain way. This ability, therefore, is likely to develop gradually in our evolutionary line. This suggests that we should partly share it with other species – the more so the closer they are to us in evolutionary terms. It would be reasonable to expect, therefore, that although our mental states might have little in common with those of an ant or a bee (if they have any), the number of common elements should increase as we move along the evolutionary tree – from invertebrates to vertebrates, then to mammals, then to primates – toward the position of our own species on it. From this viewpoint, a dog may indeed be unable to love his master – in the sense that, if we were able to feel his feelings toward his master, we would be hesitant or unwilling to call them "love." However, his feelings could be similar enough to our "love" to reliably produce certain behaviours that we would associate with love and, perhaps more importantly, to inhibit certain other behaviours that we would associate with feelings inconsistent with love, such as hatred, dominance, and so forth. Should this be the case, our attribution of "love" as a mental state to the dog would be a correct *de re* ("of the thing") attribution despite not being a correct *de dicto* ("in those very words") attribution.[85] We could, therefore, reliably predict the behaviour of the dog by analogy to ourselves and spare an explicit "theory of dog mind" that would relate its behaviour to mental concepts.

This suggests that the way to predict animal behaviour through the attribution of mental states, however anthropomorphic it may appear, can be highly effective. In any case, it should be both more effective and more natural than the body-of-propositions model. One need only know what mental states to attribute, and how. In the following sections of this chapter we shall argue that it is this knowledge that makes, for example, Nenets reindeer herders different from us, and that it is our lack of this knowledge that makes their predictions mysterious to us. First, however, a closer look at the cognitive mechanisms that can support the attribution of mental states is needed.

---

85 See Tetzlaff and Rey (2009) for a discussion of the attribution of *de re* and *de dicto* mental states to animals and their explanatory and predictive merits.

7.3 Theory of mind

It has been known for a long time that the ability to attribute mental states to intentional actors or agents plays a very important role among people. Indeed, any one person is almost constantly using this ability in interaction with the animals he or she encounters most often – fellow human beings. This ability is used for attributing emotions, intentions, and beliefs to them, predicting their reactions to our actions, and designing our strategies in interaction with them (Perner 1991). It seems that almost every instance of normal interaction between adult humans is inevitably accompanied by at least some mental-state attribution by each side to all the other sides (Carruthers and Smith 1996; Doherty 2009) – a fact that makes this interaction and perhaps even the very existence of the phenomenon of human culture possible (Enfield and Levinson 2006; Perner 1991; Tomasello 2001). Indeed, consider the following simple dialogue between a man and a woman (taken from Doherty 2009: 191):

Woman: I am leaving you.
Man: Who is he?

It is evident that the man in this dialogue has attributed certain intentions, emotions, and desires to the woman, namely that she is intending to leave him for another man, whom she probably loves. Furthermore, it is also evident that he is relying on the woman's ability to attribute to him exactly this belief about her intentions and emotions. Otherwise, his question would not make any sense to her. Finally, the very fact that we have been able to understand this dialogue suggests that we have been able to attribute the above mentioned beliefs and expectations to the man – in other words, to read his mind. Without such mind reading, the dialogue would make no sense to us.

Of course, our ability to read the minds of our fellow humans is not perfect, and we make mistakes now and then by attributing to them beliefs, intentions, emotions, and desires that they do not have. Nevertheless, this ability works well enough to provide a basis for meaningful and, most of the time, appropriate actions. It is worth noting that the human mind-reading ability has all the properties associated with a situated skill as understood by anthropologists. Indeed, most of the time, we read the minds of our fellow human beings automatically, effortlessly, with little explicit deliberation and reasoning, and in relation to concrete situations.

The enormous role played by mind reading in the everyday lives of humans has been attracting cognitive psychologists' attention to its mechanisms for more than twenty-five years. These mechanisms are usually referred to as the "Theory of Mind" (TOM), which reflects the first attempts to attribute a theoretical explanation to them. At least three theoretical models explaining these mechanisms have been proposed. The first theoretical model of mind-reading abilities – from which the very term "theory of mind" has emerged and which is currently referred to by the rather inelegant term "theory-theory of mind" (Carruthers and Smith 1996) – was proposed in 1980 by the philosopher Adam Morton. According to this model, humans "read" the minds of their fellows by making use of a general theory of the human mind's functioning, which, despite being implicit, is not dissimilar to a scientific theory (Botterill 1996). Each person develops such a theory during infancy and early childhood by observing other people's behaviour and experimenting with it (Gopnik and Wellman 1992). This theory somehow describes in general terms the way in which knowledge, emotions, and desires are acquired and processed by humans, as well as how they are manifested in their actions (see Doherty 2009; Perner 1991 for a more detailed description of this model). The second model, called the "modular theory of mind," owes its origins to Alan Leslie (1987). It currently exists in many variants whose common feature consists of the claim that mind reading is performed by a special processor or module "hardwired" into the human brain and shaped by natural selection. Since this module is a part of the genetically encoded "hardware" of the brain, our ability to read minds is innate. The variants of the modular theory differ with respect to how exactly this module works (see Doherty 2009; Nichols and Stich 2003 for details and criticism). The third model, usually referred to as the "simulation theory of mind," was proposed simultaneously and independently by Robert Gordon (1986) and Jane Heal (1986), and further developed by Paul Harris (1992). Its fundamental assumption is that people read the minds of others by simulating them using their own mind. In other words, people "put themselves into other people's shoes" by representing in their minds the conditions in which these others find themselves and the information available to them, and then "run their own mind off-line" (Harris 1996) to process this data. By so doing, they can represent the beliefs, emotions, and intentions the others are likely to have, since cognitive mechanisms are essentially similar among all humans (see Harris 1992 for details of this model).

During the last decade there have been attempts to merge the simulation theory and the theory-theory (Nichols and Stich 2003; Perner and Kühberger 2006; Perner 1996). According to these mixed models (of which the model of Nichols and Stich (2003) is to date the most developed), people can use both implicit knowledge (which may not, in this case, be so general and theory-like) and simulation for mind reading. For example, general knowledge about human perceptual abilities and emotional reactions can be used for attributing desires and some discrepant beliefs, while the attribution of inferences (reasonings) and plans is achieved by simulation (Nichols and Stich 2003).

7.4 Reading the minds of non-human agents

All the described cognitive models of mindreading have been developed to explain how humans read the minds of other humans. On the other hand, as was mentioned above, it seems to be rather natural for humans to attribute mental states to non-human agents as well. In fact, people tend to attribute agency[86] to any object that demonstrates certain specific properties or behaviour, and interpret it by making use of Theory of Mind (Susan C. Johnson 2003). Some researchers (e.g., Leslie 1996; see also Nichols and Stich 2003; Tremlin 2006) claim that humans are equipped with a special (probably "hardwired") cognitive mechanism called the Agency Detection Device (ADD), which analyses every object people encounter and marks it, using still unknown clues, as either having or lacking agency. The objects marked as agents are automatically processed using Theory of Mind; in other words, an attempt to attribute mental states to them is made. Other researchers (e.g., Doherty 2009) doubt that such a device exists. According to them, any object demonstrating a kind of behaviour that we cannot make sense of on the basis of our folk physics[87] (for example, any object moving in a complex trajectory), is automatically attributed with agency. In any case, whatever mechanisms humans use for attributing agency, animals represent the type of objects to which agency normally gets attributed. Empirical evidence of this has recently been offered by Nicola Knight (2008) in his study of mind reading among the Yukatek Maya. This study demonstrated that the ability to read animal minds and correctly predict animal behaviour correlates developmentally as well as across populations with the ability to read human minds. This study also demonstrates that mind reading, that is, the attribution of mental states to animals, is indeed used by aboriginal people to decipher and predict animal behaviour.

It is rather surprising, therefore, that very little research on humans' "reading" of animal minds and the implications of the existing Theory of Mind models to this process has been carried out in cognitive science to date. Perhaps the most prominent studies on this topic are those of Justin Barrett and his colleagues. In one of them (Barrett et al. 2001), a group of children of different ages was shown a closed saltine cracker box with a picture of crackers on it and a closed unmarked paper bag. The children were asked what they believed was in the cracker box. All of the children responded that the cracker box contained crackers. Then the box was opened and its real contents – which consisted of small rocks – were shown to the children. The children were also shown that the crackers were actually inside the bag. After re-closing both the bag and the box, the experimenters asked the children a set of paired questions in the following form: "If I showed this box to <Agent X>, what would he or she think was inside it?; If <Agent X> wanted some crackers, where (in the bag or

86 The term "agency" is used here in the way it is normally used in psychological and cognitive literature. It refers to the capacity to behave purposefully or intentionally and is, therefore, essentially a category that is either "present" or "absent": something can either have this capacity (and be an agent) or lack it (and be an object). It should not be confused with agency as a quality that can be present to differing degrees, as it is usually understood by anthropologists and sociologists.

87 The term "folk physics" refers to the body of implicit knowledge and expectations humans have about physical objects.

in the box) would he or she look for them first?" The list of agents included a human, a bear, an ant, a tree, and God. The aim was to see if the attributions and predictions made by the children would be different for different agents and, if they were, what the differences would be.

Differences in attribution were indeed observed, especially among children more than four years old. The majority of them responded that the human would think the box contained crackers and would look for the crackers first inside the box. On the other hand, the majority of children believed that God would know the crackers were inside the bag and act accordingly. The responses about the other agents differed. For example, while responses about the tree were mostly similar to those about the human, some children believed that an ant and, more frequently, a bear would know that the crackers were inside the bag. They conjectured, for example, that "a bear would look in the bag and not in the box for crackers, because it would be able to smell where the crackers were" (Barrett et al. 2001: 56). The authors concluded, therefore, that people can apply their Theory of Mind differently and selectively to different types of agents (such as those that are non-human) by taking into account their beliefs about the differences between these agents and humans. This concerns at least perceptual differences (Barrett et al. 2001: 60–63).

In a later work performed with different collaborators (Barrett et al. 2003), Justin Barrett demonstrated that children can also attune their Theory of Mind to the background knowledge that a given type of agent is likely to have. Based on these findings as well as those of the previous study described above (Barrett et al. 2001), the authors concluded that people can have many specific theories of minds to apply to different types of agents. Thus they are likely to have a separate theory of dog mind, of bear mind, and so on. However, these theories are developed to different degrees. In a case where the theory of mind of a given agent is not developed enough to allow a mental state attribution, "a more developed and more salient theory of mind, such as human theory of mind…, might fill in the inferential gap" (Barrett et al. 2003, 106). The authors further assert that, "Given that a human theory of mind will be the most complete and salient of all theories of mind, it will be drawn upon most frequently" (Barrett et al. 2003, 106). This, as Barrett and colleagues argue, explains certain regularities in mental state attribution. This explains, for example, why the tree in the first of their studies was regularly treated similar to the human but not to God.

Generally, Barrett and his colleagues use their results as evidence against the simulation model and in favour of the theory-theory model (Barrett et al. 2001: 62–63). However, their findings are also not incompatible with the mixed theory-simulation model as proposed by Nichols and Stich (2003). Indeed, it could be suggested that the beliefs about animal perception, background knowledge and, probably, emotional life are used for attributing beliefs and desires to animals, while further inferences about what the animals would "reason" and do are achieved by simulation. In other words, the role of the animal-specific theory of mind appears to be mainly to attune the simulation process by enabling the predictor to better put him or herself "in the animal's shoes," while the actual prediction is made by running the predictor's mind so attuned "off-line." In this case, the animal-specific theory of mind may be relatively undeveloped in the sense that it is probably not an actual "theory," i.e., a body of knowledge explicitly or implicitly relating a range of observable phenomena to a set of abstract theoretical concepts and specifying relationships between those concepts. It could be more justified, therefore, to use the term "attuning information" rather than "theory of animal mind" to refer to it. We suggest that this information can include:

1. some knowledge (representation) of animal perception;
2. some representation of the previous experience of animals concerning the items they perceive;
3. some representation of the animals' susceptibility to different basic emotions;
4. some information about how animals act on (or handle) these emotions.

It would appear that information related to 1 and 2 would be necessary to attune one's mind for effective simulation of the animal's beliefs; information related to 2 and 3 would be required do the same for simulating the animal's desires; and information related to 1 and 4 would be involved in simulating the animal's intentions.

This model is quite consistent with the results obtained by Barrett and colleagues in their studies. Indeed, it was exactly the information related to 1 and 2 that the children in these studies were observed to make use of in predicting the behaviour of specific actors. Furthermore, similar to Barrett's "multiple theories" model, this model would also explain why the tree was treated identically to the human but not to God or the bear: having no information about the mental life of trees, the children could not attune their simulation specifically to the tree and had to simulate it using their own, that is, specifically human, beliefs and desires (the phenomenon of "default belief attribution" as Nichols and Stich (2003) have called it). Furthermore, this model has an advantage because it does not postulate the complex procedure of filling the gaps in one theory of mind by taking parts of the other theory.

However, there is an even better reason to think that simulation is involved in attributing mental states to animals. Let us recall once again the explanation provided by the Nenets herder for why he thought the reindeer would come to the camp by themselves the following morning: "The mosquitoes will harass the herd. The reindeer will want to get a break from them and come to the camp by themselves… because they will reason that the smoke from our fires will drive the mosquitoes away." As is clear from the discussion thus far, this explanation is best understood not as an account of propositional rules of reindeer behaviour expressed in the language of anthropomorphism (as the older generation of cognitive anthropologists would probably treat it), nor as a *post-factum* anthropomorphic reflection on implicit and situated skill, the exact mechanisms of which are unknown to the informant himself (as Ingold, Lave, and other theorists of the knowledge-as-skill model would probably argue). Rather, the informant has provided exactly what he was asked to provide – the justification of his conclusion about the future behaviour of reindeer framed as a set of attributions of mental states he has made in order to reach that conclusion. No person other than the naive anthropologists lacking a clue as to the type of thinking involved would need or expect anything more. However, this explanation is still anthropomorphic in the sense that the mental states the informant has attributed to reindeer were human *emotions*, human *desires*, and human *beliefs*.

Now, as has been argued earlier in this chapter, predicting animal behaviour through the attribution of human-like mental states to the animal can be effective. The animals' intentionality *can* be processed by the human mind *in terms of* human emotions, desires, and beliefs taken either as purely theoretical concepts or as *de re* rather than *de dicto* descriptions of mental states the animal indeed possesses. However, it need *not necessarily be* processed in this way if the general "theory of reindeer mind" (in the sense of the theory-theory model) constructed from experience is the only thing involved in this processing. Indeed, such a theory can be constructed in any terms and, furthermore, constructing it in less anthropomorphic terms could easily increase its efficiency. The fact that Nenets herders as well as all other people, aboriginal or otherwise, described in the literature (including, by the way, animal psychologists – see, e.g., Cheney and Seyfarth 1992; deWaal 2006: 59–67) use anthropomorphic emotions, desires, and beliefs to attribute mental states to animals suggests that the human mind is probably simply unable to attribute anything else. This, in our opinion, rather clearly indicates that at least some simulation (for which, of course, the human mind working only in terms of human desires, emotions, and beliefs is used) is involved.

Of course, all other things being equal, a correct prediction of animal behaviour through a theory-simulation mix is possible only if the computations performed in the animal's mind are at least to some degree isomorphic to those that are normally performed in the human mind. This is necessary in order for the latter to serve as a satisfactory model of the former. As was mentioned earlier in this chapter, there are good evolutionary grounds for believing that this condition can be reasonably satisfied as far as higher animals (presumably including reindeer) are concerned. Indeed, there is a long tradition in science of using animal models for studying processes that take place in both the human brain and the human body. This tradition is well established, justified on evolutionary grounds, and produces significant results. It would indeed be strange to claim, therefore, that the converse, that is, using the human brain and body as a model for processes going on in animals, should be absolutely impossible or unfruitful. However, as far as interpreting and predicting reindeer behaviour by reindeer herders is concerned, there are reasons to believe that simulation can be successful even if the computation procedures used by the herders and the animals are radically different. This is so because in this particular case, all other things are not equal. As we argued in detail in the previous chapter, the reindeer behaviour that the herders have to interpret and explain emerges from the process of dynamic mutual adaptation of reindeer and herders to each other's activities. It can be expected, therefore, that these patterns represent the most effective (most rewarding) courses of actions in the situations that the herders and the reindeer together define and participate in. Since the herders actively (although probably not entirely consciously) participate in negotiating the essential parameters of these situations with their animals, they can easily infer, using their own computational abilities, the courses of actions that are likely to be the most rewarding for their animals. This would yield the correct predictions of the animal behaviour even if it results from an entirely different inferential pattern.

The following example illustrates more clearly what we have in mind here. As was mentioned in the previous chapter, Nenets reindeer tend to move to the campsite if attacked by predators or some critical mass of mosquitoes. It can be supposed that the Nenets herder that we quote in this chapter alluded to exactly this sustainable behavioural pattern when he predicted that "the mosquitoes will harass the herd. The reindeer will want to get a break from them and come to the camp by themselves." Now, it may well be that the mental process he attributed to the animals as the basis for this behaviour ("The reindeer will reason that the smoke from our fires will drive the mosquitoes away. And they will know that our campsite is the nearest place without mosquitoes") does not adequately simulate what is really going on in the animals' brains. It is not inconceivable, for example, that this particular behavioural pattern is based on conditioned learning: being rounded up and driven to the campsite daily, the reindeer could simply learn by reinforcement the association between herders' nomadic tents and fires on the one hand, and relief from mosquitoes on the other. Therefore, rather than being engaged in a complex computation (reasoning) involving representations, they could simply follow the learned reaction. However, in our opinion, this does not matter. Indeed, at the end of the day, the relief from mosquitoes near the campsite indeed occurs due to the smoke of the fires made by the herders, and the reindeer could learn to come to the campsite exactly because of that. Therefore, the reasoning the herder has attributed to the reindeer is a good *de re* description of the logic behind their behaviour. Thanks to the dynamic mutual adaptation, the reindeer can be expected to follow this logic even if it does not accurately reflect their mental processes.

In a sense, the situation here is not dissimilar to that of playing chess with a computer, as famously described by Dennett (1981). Similar to playing chess against a human player, playing with a computer involves constant attempts on our side to predict the moves the machine could possibly make. These predictions almost always take the form of attributions of mental processes. For example, we might think something like, "If I move my bishop there, the computer could take it with its knight. But it will definitely not take my bishop with its knight because it will reason that, in that case, I will take its queen with my rook on my next move." We would then decide on our move (for example, to move our bishop to the place where it can be taken by the computer's knight) on the basis of this attribution. If we are moderately good chess players, our predictions based on such attributions of mental processes will tend to be correct most of the time. Now, nobody would probably seriously think that the mental processes that we attribute to the computer even remotely resemble what is actually going on inside the machine. Indeed, the algorithm of the chess program is normally based on very different principles than those guiding our inferences, and it cannot be adequately simulated in our mind. Still, predicting the computer's moves through simulation and attribution of mental processes does work effectively for the following reasons: 1) we can represent the situation in which the computer makes its moves (including the range of factors that limit its course of action), and we can assume that the computer also perceives this situation and these factors (in other words, we know the rules of chess and the current position of figures on the board, and we can assume that the computer also knows them); 2) we can safely assume that the aim of the computer is to win; 3) we can safely assume that the design of the computer program, whatever it is, allows the machine to pick up the most effective (from the viewpoint of its aim) course of action in the given situation (see Dennett 1981 for a further discussion). This allows us to use our own mental capacities to infer what would be the most logical move for the computer to perform. Just as in the case of reindeer, the computer can be expected to perform this most logical move regardless of the computational process it actually used to infer it.

Obviously, we can fail to correctly predict the moves of the chess-playing computer, and there are two principle reasons for this. Most obviously, we can fail if the computer program is able to pick up a more effective move in the given situation than we can infer using our mental processes. It can happen, for example, that taking our bishop and allowing us to take the queen in the next move is an effective move, because it would allow the computer to win afterwards. More importantly, however, our predictions can fail if one or more of the three assumptions described in the previous paragraph are false. In other words, the computer can behave unexpectedly if a) its program is based on a different set of chess rules than those we know; b) if the program is not designed to win; or c) if the program is so bad that the moves it makes are manifestly stupid, that is, not at all the most effective in the given situation (as any chess player well knows, stupid moves can still be harmful to the opponent exactly because they are unpredictable). In the case of herders predicting reindeer behaviour, the analogue of such a flawed program would be a "wild" reindeer, that is, the reindeer whose behaviour has not been adapted to the given herding system through the process of dynamic mutual adaptation.[88] As noted in the previous chapter, herders do have problems with such animals and normally lose them quite quickly, for which there are technological reasons. Now we can add that there are likely to be cognitive reasons as well.

88 Recall that such an animal is not necessarily wild from our point of view. Rather, he may be considered wild because his behaviour has been adapted to a herding system different from that of the herders attempting to make sense of the behaviour. The animals transferred from one herding system to another (briefly mentioned in the previous chapter) are an example of this.

7.5 How can one learn to think like a reindeer?

Stories and fairy tales about animals constitute an important part of the body of Nenets folklore. A researcher living with Nenets will hear two or three animal fairy tales every week without having asked for them, and this number increases greatly if he or she expresses interest in animal behaviour. It is simply normal for Nenets herders to respond with a fairy tale or an anecdote when asked how a particular animal behaves.

It has been claimed (Pushkareva 2003) that fairy tales contain traditional knowledge and norms (whether ecological or social), and that telling them plays an important role in the transmission of this knowledge. We are inclined to believe this. However, let us consider a typical fairy tale about a hare that we recorded in spring 2007:

> A hare is jumping along the path he has previously trodden around bushes on the bank of a river. A polar owl comes and says to the hare, "Hare, do not continue along this way because a hunter has put a trap on the trodden path." The hare escapes from the path and tries to jump through virgin snow. However, his legs keep getting stuck in the snow and he starts to worry about how slowly he is moving. "I would like to return to the path," says the hare, "because otherwise I will not get to the place I need to be for ages." He returns to the path and gets trapped immediately. While he is sitting in the trap, the owl returns and says, "Did I not tell you about the trap?" "Yes," says the hare, "and I regret that I did not do as you suggested." (Ivan Lamdo (Shurshakov), 54, April 2007)

What kind of traditional knowledge does this fairy tale communicate? Of course, one could suggest that it simply teaches that hare traps should be placed on the paths the hares tread in the snow.[89] In fact, the informant who told us this fairy tale said it is often interpreted that way by young hunters. However, he insisted that it is a poor interpretation: "If the old people wanted to teach us how to place traps, they would simply say, 'Put traps on a hare path.' They told a fairy tale because they wanted to teach us about hares, not about traps." In light of our earlier discussion, this remark makes perfect sense. Indeed, it is rather easy to see that the fairy tale contains first and foremost information on how the hare experiences the snow cover. Furthermore, this "attuning information" is exactly what a person needs to interpret and predict the behaviour of hares by means of a hybrid (theory/simulation) process of mental states attribution. Indeed, the fairy tale explicitly teaches the listener 1) how hares perceive the snow cover; 2) what their previous experience of snow cover is; 3) what emotions they are susceptible to in relation to the snow cover; and 4) how they act on these emotions (cf. the discussion of "attuning information" and its parts in the previous section of this chapter). In this sense, a listener indeed can learn a lot about hares from

89 One could also suggest something more abstract, for example, that the fairy tale teaches one to listen to the advice of more experienced people. However, such an interpretation does not correspond at all to the situations in which such fairy tales are usually told and, therefore, is not considered here.

this fairy tale. In order to do that, however, he or she needs to know what bits of the fairy tale require attention and how the acquired information should be used.

In contrast to the Nenets, Komi reindeer herders of the Bol'shezemel'skaya tundra do not tell animal fairy tales unless they are prompted to do so. Their way of transmitting information about animal perceptions and feelings is much more direct and situated, as the following example illustrates. In August 2003 one of us accompanied Alexei Filippov, a 56-year-old Komi reindeer herder, on his 24-hour shift with the herd. After the herd was rounded up and set on its track across a chosen piece of pastureland, the herder and the anthropologist took up a position on a nearby hill from where the herd could be easily observed. There the informant said that, as the anthropologist was interested in learning about reindeer, he would show him something. He continued:

> Reindeer feel good when they have at least two friends in front of them, two on each side and two behind. Look how this one [pointing to a reindeer that had been left behind as the herd moved forward] is eating with his head down to the ground. He is eating, but he is also listening to how – *krup-krup, krup-krup* [imitating the sound of reindeer chewing] – his friends are eating nearby. But now – oh – there is no *krup-krup*! He'd better raise his head! [The reindeer raises his head and rushes toward the center of the herd]. Oops! I have been left behind! How frightening! I need to run!

Such situated commentary on reindeer behaviour can be encountered rather frequently among Komi. More importantly, the commentary most frequently follows a standard scenario: first a single animal or a group of animals is pointed out. Then its perception of the situation and beliefs about it are made explicit. The animals' emotional reaction to the situation is orally reconstructed on that basis. Then the novice is left to observe for himself how the emotions just reconstructed result in actions. Like the animal fairy tales of the Nenets, these in-situ commentaries of the Komi explain exactly the types of information that are needed for attuning the mental simulation to interpret and predict animal behaviour.

Can we conclude, therefore, that both types of narratives described are instruments for transmitting the theory of reindeer mind as a culturally specific skill? Our reply is, "Yes and no." Yes, because the information essential for executing this skill is certainly transferred. No, because they, of course, do not transmit the skill itself. Indeed, as the Nenets informant mentioned, inexperienced hunters would take from the fairy tale a lesson about traps rather than about hares. Similarly, the only wisdom we could acquire from the comments of the Komi informant at that particular moment was that reindeer always tend to move from the periphery to the center of the herd – a message that would have served us poorly, by the way, as we recognised from our own herding experience a few weeks later that there are countless situations where they do not. As should be clear from the above discussion, the information made explicit in the comments and fairy tales is essential, but only if it is used in a special way: to attribute mental states to the animals. This way of using the information is not – and hardly can be – taught. It can only be discovered from the experience of applying the information to interpret and predict animal behaviour. In other words, the theory of animal mind – just like the theory of human mind – is acquired by what Tim Ingold aptly describes as "guided discovery" (Ingold 2000).

This fact is essential for understanding how the skills of predicting and explaining animal behaviour are likely to be distributed in society. Indeed, it appears that novices as well as people with limited interaction with animals (including anthropologists) tend to lack this skill, even if they have acquired all or part of the information needed for its execution. They do not attribute mental states to the animals and tend to reinterpret the relevant information they possess as propositions and rules of thumb not dissimilar to those suggested by the older approach to traditional knowledge. As another Komi informant nicely put it:

> All *iando* [novices, inexperienced people] are like that. They hear others saying, "Reindeer like this; reindeer don't like that," and then they expect reindeer always to do this and not that. When the reindeer do not behave as you expected, you come and ask why. Reindeer also have their heads, you know, and their heads are not very small! That is why! In order to become a good herder one should learn to think like a reindeer!

One learns to think like a reindeer from experience with reindeer. As this experience grows, one tends to start making more and more attributions of mental states and becomes less dependent on propositions, while the propositions themselves get reinterpreted as just a type of clue to animal perception, beliefs, and emotions. On the basis of our observations and analysis, therefore, we suggest that traditional knowledge about animals is distributed throughout a society as a whole in the form of a continuum, with pure propositions at one extreme and pure mental state attribution at the other. The form of knowledge each particular individual has falls between these two extremes, its relative closeness to each of these extremes being dependent on the experience one has with the animals.

7.6 Conclusions

When Jane Lave says that everyday human cognition, including presumably interpreting and predicting animal behaviour on day-to-day basis, is essentially situated (Lave 1988), we can agree with her as long as she means that this cognition is not based on general propositions or rules-of-thumb. We agree to a certain point also with Tim Ingold when he says that a significant part of everyday cognition is skill-like and, therefore, implicit, effortless, and based on the "education of attention" (Ingold 2000). We depart from their positions, however, if they are taken to mean that there exist no high-level cognitive mechanisms that are based on abstract representations and are structurally independent of the particular situations of their execution, or that such mechanisms and processes are not activated in skillful actions. Thinking and acting, cognition and doing, mind and body may well be inseparable from one another. They may be two sides or aspects of the same thing. However, this is far from insisting that they *are* the same thing.

The skill of interpreting and predicting animal behaviour is a clear example of this. In this chapter we have tried to demonstrate that this skill is indeed implicit and situated, as well as being neither consciously reflected upon nor based on consciously held general-purpose propositions. It is highly likely that this skill is acquired from experience by means of Ingoldian guided discovery, and that it is based on educated attention to the situation and the behaviour of animals. However, the essence of this skill still consists of the purely cognitive or mental act of attributing mental states to animals based on certain attuning information, and inferring their future behaviour on this basis. It is very likely that simulation is involved as part of this skill's mechanism. It is solely this mental act, performed in a certain situation but structurally independent of it, that makes interpretation and prediction possible.

If our suggestions are correct, then the specifics of this skill in different societies can be meaningfully described and analysed using a set of qualitative and quantitative methods. First of all, a study on animal mind reading in a particular society should include the collection and recording of the attuning information used in the process. A good way to start such a collection would be an inquiry into (by interviewing experienced mind readers) and the compilation of a comprehensive list of the kinds of emotions and beliefs mind readers attribute to animals. The second step would be to elicit from informants their understanding of animal perception and its relation to the emotions and beliefs: What can particular animals see, hear, smell? What do they feel about these perceptions? Finally, the informants could be asked to produce narratives about the usual reactions of particular animals when they are, for example, afraid, or in a rage, or hungry.

Secondly, a qualitative study of the actual process of attributing mental states could be undertaken. Such a study is especially important because it would give an idea of the distribution of animal knowledge in a society along the proposition–attribution continuum. In the case of northeastern European and Siberian reindeer herding nomads, for example, such a study could include accompanying the reindeer herders during their herding activities. By observing these activities, the researcher can note which of the herders' actions toward the animals could be classified as reactions based on their knowledge of earlier reindeer behaviour. After having noted such an action, the researcher can ask the herders to recall and describe the chain of conclusions or inferences that led to their decision to perform the action, as well as its expected outcome. The narratives can later be analysed for coherence, as well as for the presence and role of propositions and mentalist concepts.[90]

90 By "mentalist concepts" we mean those concepts that directly refer to the mental states of the animals. For example, in his narrative a herder might say, "The animals got scared," or "The

Finally, cognitive experiments could be employed to cross-check and flesh out the qualitative descriptions. Such experiments could include showing photos and short video clips (or simply giving verbal descriptions) of animal behaviour or herding situations, and asking informants to interpret these situations and predict further behaviour of the animals. Specific information that the researcher (based on previous qualitative research) hypothesises should trigger or enable the attribution of particular mental states could be either present in or absent from the photos, video clips, and descriptions. The responses of informants could be compared to those predicted by the hypothesis, thereby proving or disproving it, or leading to a more comprehensive reformulation of the hypothesis.

We suggest that this methodology could produce a reliable description of the specifics of animal mind reading in a particular society. This description could be used later for purposes of education, making use of local knowledge in decision making, and better understanding the specific problems local people encounter in their efforts to adapt to industrialised society. It is also likely that data obtained in this way could be of use to practitioners of other scientific disciplines such as ethology and animal psychology.

---

reindeer decided to seek relief from mosquitoes by going up to the top of the hill."

8 CONCLUSION: CULTURE, ECONOMY, AND COGNITION

8.1 Cognitive processes as cultural phenomena

Observations presented in the previous chapters demonstrate a number of similarities as well as differences in cognition both between the two reindeer herding groups in our study, and between the reindeer herders and sedentary groups living in the same region. In this final chapter we will attempt to analyse these differences and similarities and discuss how they can enrich our understanding of the dynamic relationships that exist among the culture–economy–cognition triad. Let us start by returning briefly to the fundamental question that we posed at the very beginning of this book: How pronounced can the variations that culture plays on universal cognitive processes be? In other words, how far beyond mere cognitive content can the cultural influence on cognition go?

As was discussed in detail in chapter 1, four basic ways in which culture can play variations on cognitive universals have been proposed:

1. by structuring the naturally existing individual variation along one or several dimensions of personality (the "personality dimension" hypothesis);
2. by providing models, blueprints, or scripts that specify situations and ways in which particular universal cognitive processes ("cognitive tools") are deployed (the "sociocultural models" hypothesis);
3. by affecting the design and level of development of particular cognitive tools themselves (the "cognitive toolkit" hypothesis);
4. and by shaping settings and characteristics of situated cognitive processes (the "situated cognition" hypothesis).

We suggest that the influence of a particular culture on the cognition of its representatives channeled through each of these four types of influence would produce a specific cluster of cognitive particularities that would differ from those of the representatives of other cultures. The differences in cognitive *processes* (as opposed to the *contents* of cognition) observed between cultural groups should, therefore, fall into four broad categories corresponding to the particular channel of cultural influence on cognition. These are 1) differences in personality dimensions; 2) differences in models of thinking (i.e., in recipes specifying the application of universal cognitive tools); 3) differences in cognitive toolkits; and 4) differences in situated cognition. The mechanisms leading to the development of cultural variation in cognition, as well as the range of this variation, can be expected to differ depending on the category.

Most of the cognitive particularities described in the previous chapters can be comfortably slotted into the second and the third categories outlined above: they can be classified either as differences in cultural models of thinking, or as differences in cognitive toolkits. The differences in microlandscape typologisations and some of the differences in navigation methods between the two reindeer herding groups described in chapters 4 and 5 would belong to the former category. On the other hand, the differences in the mechanisms and forms of spatial representations between the groups of reindeer herders and between the herders and the settled groups described in chapters 3 and 4 can be treated as those involving the specific design of tools within the cognitive toolkits. The same, in our opinion, can be said of the mechanisms of deciphering and predicting reindeer behaviour. These also go beyond the level of the mere application of universal cognitive tools to the level of the presence or absence of particular cognitive tools themselves, at least as far as the formation of the "theory of reindeer mind" as a specific cognitive procedure based on a theory/simulation mix (as suggested in chapter 7) is concerned.

This latter category of differences can be a good illustration of both the depth and the limits of the cultural impact on human cognition. On one hand, the cultural variation that emerges from our observations and discussion is both deep and broad. It concerns cognitive processes involved in different domains of cognition and in several of the most important components of the cognitive cycle: higher-order perception (e.g., the perception of distance); representation (e.g., the type and structure of spatial knowledge); and inference (e.g., course setting and the prediction of future behaviour). The resulting culturally specific cognitive tools radically differ as far as their computational roles and capacities are concerned: compare the spatial representations (mental maps) of the Komi, highly coherent and oriented with respect to cardinal directions, to those of the Nenets, which are regionalised and seem to lack any allocentric orientation. As a consequence, the behavioural patterns supported by these tools also differ in important ways. Nevertheless, all these tools are still recognisable variations of well-known cognitive universals: they differ only in their details (albeit quite important ones) rather than in their underlying principles. In our study of reindeer herders' cognition, we have not observed any behaviour or cognitive capacity that could not be explained using one of the models of cognitive processes already developed in cognitive science. We are not, therefore, suggesting the existence of an entirely new cognitive process that has not been observed in representatives of other groups and cultures (including individuals from modern industrialised nations).

This fact, in our opinion, is very much consistent with – and indeed provides additional support to – the doctrine of psychological universalism. Culture can influence cognition to a very significant degree by privileging some cognitive processes over others both in development and use, and even by modifying details of these processes. This can result in behavioural or mental abilities that, at first glance, can appear mystical or superhuman to representatives of other cultures (see, for example, the prediction of reindeer behaviour described in the first section of chapter 7). However, culture cannot create completely novel cognitive processes that qualitatively differ from those present in representatives of other cultures. Nor, it seems, can culture completely eliminate any of the universal cognitive processes, creating societies whose representatives would lack particular behavioural abilities or cognitive capacities present in other cultures. Returning to the example of three-dimensional vision mentioned in chapter 1, we may well encounter groups or societies whose perception of visual perspective differs from our own. For example, they may use a different set of depth cues, be vulnerable to different classes of visual illusions, and so forth. However, we are not likely to encounter either a society whose representatives are unable to perceive the world in 3-D or a society whose members perceive the world in a set of perceptual dimensions that are radically different from the dimensions of height, width, and depth.

Despite the clear limits of culture's influence on cognition, the range of cross-cultural variation in cognition within these limits is still remarkable. This raises an important question about the mechanisms of this influence: How exactly it is realised, and why do particular cognitive variants result from it? In the rest of this chapter, we shall draw on the evidence presented in the earlier chapters to suggest some answers (necessarily partial) to this question.

8.2 The concept of cognitive specialisation

For a long time, the discussion of the mechanisms of cultural influence on cognition in both anthropology and psychology focused on two main topics: socialisation practices and conceptual (categorisation) systems. The former topic was introduced to the disciplines mainly through the seminal (although recently criticized) works of Margaret Mead (1928, 1930). Following the publication of these works, it quickly became commonly accepted that culturally specific practices of rearing and educating children have an enormous formative effect on their developing cognition mainly (but not exclusively) through encouraging particular emotions, judgments, and behaviours, and discouraging or strictly prohibiting others. This has been believed to lead to the formation of sustainable and culturally specific patterns of emotion, motivation, and thinking. This formative effect, in turn, is commonly believed to represent the most important mechanism in the creation of cross-cultural psychological diversity (Barry et al. 1959; Benedict 1934; Berry 1976; Markus and Kitayama 1991; Witkin 1967). Even the demonstration that the range of cross-cultural diversity in socialisation practices had been overestimated (among others, by Mead herself – see Freeman 1983; Holmes 1988) did not seem to have a significant effect on this belief, which is still widely accepted (see, e.g., Nisbett 2003).

The second topic – the conceptual or categorisation systems – was introduced in the 1960s and is often connected to the works of Claude Lévi-Strauss (e.g., Levi-Strauss 1966) and the so-called structural anthropology school he founded, but most of the research on this topic has been conducted since the 1970s in the framework of cognitive anthropology (D'Andrade 1995). The central argument in these studies has been that conceptual systems – the culturally specific structured systems of categories or concepts that a person learns as part of acquiring his or her native language as well as from engaging in different kinds of practical activities – create structures that allow the person "to turn the natural [perceptual? – *K I, M.D.*] world into things and to think of the things so identified in terms of their essential qualities" (Ellen 2003, 43). In doing so, these structures have an essential impact on cognition by structuring higher-order perception, memory, and inference in culturally specific ways. As was noted in chapter 1, this argument plays a particularly important role in the "sociocultural models" approach, where conceptual systems represent the most frequently researched type of sociocultural models to date. In the framework of this approach, studies of how conceptual systems employed by members of particular cultures and societies are rooted in their languages ("folk categorisations" or "folk taxonomies") and practical activities have predominated (see, e.g., Atran 1990; Berlin 1992; Ellen 2006, to name just a few). On the other hand, the whole approach has recently been criticised, particularly by the representatives of the situated cognition school (e.g., Ingold 2000, 2003; Lave 1988, 1996).

It is not particularly surprising that socialisation practices and conceptual systems have attracted a disproportionally large share of attention from scholars interested in mechanisms of cultural influence on cognition. Indeed, both phenomena are arguably unique in two respects: on the one hand, they are manifestly cultural in the sense that they are directly related to and determined by such "classic" cultural phenomena as religious beliefs, social systems, roles, norms, and so on. On the other hand, their potential impact on cognition is undeniable. Therefore, both represent channels through which cultural phenomena can have a direct impact on the human mind. What is probably more important, however, is that both represent particularly good *post-hoc* explanations of cultural cognitive phenomena. Indeed, it is very difficult to grasp all aspects of a conceptual system and, particularly, of a certain set of socialisation practices that could potentially have cognitive effects. Furthermore, it is practically impossible to envisage what exactly these effects might be, especially when considered in relation to other relevant aspects of the same socialization practices and/or conceptual system. Therefore, any cognitive particularity detected among representatives of a certain culture can potentially be explained *ad hoc* as the effect of some aspects of their conceptual system or their child-rearing practices. In a sense, saying that a certain people have a particular cognitive ability or particularity because they "were raised that way" is not very different from saying that "they were born that way": both statements are irrefutable (in the last instance, any aspect of cognition must undergo a certain process of biological development), but they really do not explain much. They are handy, however, if the aim is to avoid explanation.

This is not to say that socialisation practices and/or conceptual systems cannot be important mechanisms of cultural influence on cognition: they can and most certainly are. Rather, our argument is that before searching for particularities of child rearing or a set of exotic concepts that can explain certain culturally specific cognitive phenomena (this quest is with all likelihood going to be successful and, therefore, can be safely postponed), it makes sense to consider other possible mechanisms. Although these mechanisms may be less direct, they potentially offer a more reliable explanation of how culture produces cognitive effects.

Two recent theoretical developments suggest intriguing possibilities for what such indirect mechanisms might be. One of them is the new treatment of the relation between the grammatical structures of language and cognition that has been developed most notably by Stephen Levinson (2003) and his colleagues (e.g., Haviland 2000). Of course, these researchers are not trying to reanimate extreme versions of the infamous Sapir-Whorf hypothesis, which posits that the grammatical structures of a language directly determine the cognition of its speakers (i.e., that the representations a person can form and the operations he or she can execute on them are both determined by and limited to the grammatical categories of concepts and the grammatical rules of relating those concepts to each other that are present in his or her language). This extreme version of linguistic determinism was rejected both in linguistics and in psychology a long time ago. However, the authors suggest that grammatical structures can still affect cognition indirectly by priming the speaker's attention to and ensuring his or her representation of certain aspects and distinctions of the natural and social environments, to the neglect of others. For example, in some languages a verb must take a form that reflects how the speaker has learned about the action he or she is referring to.[91] Thus, different verbal forms have to be used to refer to actions the speaker has witnessed directly, inferred from direct evidence, inferred from indirect evidence, was told about by others, concluded logically but without any evidence or information from others that it must have taken place, and so on (see Aikhenvald 2006 for an extended discussion). Certainly, a speaker of such a language should constantly pay attention to – and keep a record of – the sources of the information he or she communicates to others. This constant representation of the epistemological sources of information would make his or her cognition different from that of a speaker of English or any other European language.

The linguistic representation of spatial relations, which was the main focus of Levinson and his colleagues' research and has already been discussed in some detail in chapter 3, is another example of the indirect influence of language on cognition. This

91 This is known as "grammatical evidentiality." The Tucanoan and Nambikwaran groups of languages that are spoken by some indigenous peoples in Amazonia are probably the most striking and well-known examples. See Aikhenvald 2006 for details.

work, moreover, takes us all the way to the level of cognitive processes (the actual composition of one's cognitive toolkit). To recall it briefly, the researchers point out that in quite a number of languages references to spatial relations are based not on the egocentric or relative categories of "left," "right," "back," and "front," but rather on allocentric or absolute categories such as "north," "south," "east," and "west," or their culturally specific equivalents. This kind of reference to spatial relations is possible only if the speaker has a well-developed representation of the positions of objects in relation to each other and to the allocentric axes, that is, an allocentrically oriented mental map. Furthermore, the speaker has to constantly fix his or her position in relation to both the axes and the objects (in order to be able to say, as he or she would be expected to do, that something is situated, for example, "to the north of here"). In other words, the speaker should develop and constantly employ cognitive processes that many speakers of European languages do not need and, as recent research indicates (e.g., Bohbot et al. 2004), often fail to develop.

The second recent theoretical development that illuminates the possible mechanisms of culture's influence on cognition is the notion of "guided discovery," introduced by Tim Ingold (2000). The main and quite obvious idea behind this notion is that the kinds of immediate situations the representatives of different cultures would routinely encounter differ significantly. Indeed, the situation of being in the tundra far away from any kind of shelter or settlement and in urgent need of finding a way home before nightfall is something a reindeer herder can expect to encounter much more often than an office worker in a major European city. The mere frequency and routine of experiencing particular situations can be expected to lead the individual to the discovery of mechanisms that enable effective behaviour in just those situations. This is particularly true if the individual encounters such situations in the company of more experienced persons, whose behaviour the individual can observe and try to copy. As a Gibsonian, Ingold describes this discovery primarily as a heightened sensitivity to affordances contained in the situations (the "education of attention"), which enables skillful behaviour in those contexts. However, nothing in his argument excludes the possibility that higher-order, less content-specific cognitive processes, such as certain inference or representation models, could also be discovered in this way. Furthermore, as the situations encountered by members of a particular society or cultural group are bound have some degree of similarity, and since these situations are often experienced by groups of individuals, the resulting cognitive processes are likely to be, to a certain degree, similar among the members of the group even if they have been discovered by each member individually.

The most important point we want to make about these theoretical approaches is that they explain how certain cognitive particularities can get shared among the members of a given society or cultural group without being part of their shared conventions, norms, scripts, or any other collective representations that constitute culture in the strict sense of this word. Indeed, while a language with an absolute frame of reference to spatial relations can be regarded as an intersubjective entity shared by the members of the corresponding community, the oriented mental maps and the processes of position fixing do not constitute a part of this language. A linguist who has learned the vocabulary and grammar of such a language is not likely to become an expert mental mapper and position fixer as a result. On the other hand, as the examples of the two reindeer herding groups described in this book (as well as that of expert European mariners) would attest, one can become an expert mental mapper and navigator without mastering such a language. The mental mapping and position fixing processes get developed as tools for solving particular tasks that routinely arise in the course of particular activities performed in a specific situation: speaking an "absolute" language and having to make a spatial reference; being on duty with a reindeer herd and having to figure out the course of future herding actions; being alone on a sailboat and having to sail to a particular port, and so on. In all of these cases, culture plays an enormously important role in determining the kind and details of the activity and of the situations in which it is performed. By so doing it largely specifies both the tasks that need to be solved and the settings in which they are encountered. However, the cognitive particularities get developed as each individual is confronted with and must respond to these tasks and settings; they are not provided by culture as such. It can be said that these cognitive particularities result from each individual's specialisation in solving the particular tasks in the particular settings, both of which are specified by his or her culture.

In light of the preceding discussion, we would like to propose the term "cognitive specialisation" (in the singular) to refer to the mechanism whereby culture shapes individual cognition, and "cognitive specialisations" (in the plural) to refer to the cognitive particularities it produces by means of this mechanism. We suggest, furthermore, that cognitive specialisation so understood plays a very important role in inducing cross-cultural variation in cognition, particularly as far as the variation in "cognitive tools" is concerned. Indeed, while the models guiding the application of these tools may be collectively shared and, therefore, directly learned by an individual from communication with others and/or from the observation of the behaviour of others, the tools themselves cannot. They are tools for creating and manipulating symbolic representations, which means that they themselves have to remain outside the sphere of such representations. On the other hand, only symbolic representations can be collectively shared. This means that culture can create them only by indirect means, and cognitive specialisation seems to be the most obvious of these means.

8.3 Cognitive specialisations of Komi and Nenets nomadic reindeer herders

Many of the cognitive particularities that we detected among the reindeer herders represent obvious cognitive specialisations. Furthermore, in many instances the analysis of these particularities suggests particular kinds of tasks that most probably caused their development and, therefore, illuminates particular aspects of cognitive specialisation as a mechanism of cultural influence on cognition. Most of the cognitive particularities described in the previous chapters can be divided into three groups: 1) cognitive specialisations that address the tasks posed by nomadic reindeer herding as a specific *way of life*; 2) cognitive specialisations that address the tasks posed by a concrete *way of practicing* nomadic reindeer herding, that is, by a specific reindeer herding *technology*; and 3) cognitive specialisations that address tasks that are not directly related to reindeer herding as such, but that have to be solved in the specific settings created by reindeer herding as a way of life or as a specific economic practice.

The first group includes those cognitive specialisations that respond to the tasks posed by nomadic reindeer herding irrespective of the concrete specifics of the reindeer herding practice or other aspects of culture. Therefore, one can expect to find these specialisations among members of any group of nomadic reindeer herders. At least two specific aspects of cognition described in the previous chapters would fall into this group: the heavy reliance on mental maps as a form of spatial representation, and the systematic use of mental state attribution to account for and predict animal behaviour. Indeed, these two aspects of cognition have been observed among representatives of both reindeer herding groups in our study, while many representatives of other groups – including Nenets fishermen who share the majority of cultural traits outside the domain of reindeer herding with the Nenets herders – lacked one or both of them. The special status of these two cognitive aspects is not difficult to explain. Indeed, it seems rather obvious that predicting and deciphering reindeer behaviour and learning and representing the spatial distribution of those elements of the natural environment that affect reindeer most (particularly those areas with certain types of vegetation and microlandscape) are tasks that any reindeer herder would face independent of the details of his or her reindeer herding practice. As we argued in the previous chapter, the attribution of mental states seems to be the most effective way of deciphering and predicting behaviour of higher animals. Similarly, the most effective and inference-rich way to represent a spatial distribution of areas is through the integration of 2-D-S elements into the cognitive map by the means of mental mapping.

On the other hand, the details of a given reindeer herding practice can affect these two basic cognitive aspects by specifying the details of the cognitive tasks the herders have to solve on a day-to-day basis. These would produce the "second-order" cognitive specialisations that would fall into the second of the groups defined above. Three examples of such specialisations have been described and explained in the previous chapters. The first is the regionalised and multi-layered character of mental maps among the Nenets, as opposed to the coherent mental maps of the Komi that seem to have only one layer and cover a rather restricted expanse of territory on both sides of the migration track. Indeed, it is logical to suggest that this difference in the structure of mental maps is directly related to the difference in the land use patterns of the two groups: Komi, who have established migration routes and pasture their herds near them (see chapter 2), regularly face practical tasks for which the detailed knowledge of the territory directly adjacent to their long migration paths is of overwhelming importance. They rarely encounter tasks that would demand a detailed knowledge of extensive areas of land far away from the migration path. Nenets, on the other hand, do not have established migration routes and regularly change the areas of their annual migration (see chapter 2). Therefore, their everyday practice would demand a detailed knowledge of certain restricted areas they use for pasturing their herds, as well as of the locations of these areas in relation to each other, but less detailed knowledge of the rest of the tundra that they do not use for pasturing reindeer. In chapter 4 we argued that this difference in the structure of mental maps makes possible further differences in the spatial representations of the two groups, namely representing points in relation to allocentric axes among the Komi in contrast to the representation of angles and distances between points among the Nenets, as well as in procedures of position fixing and course taking. The reader should be reminded, however, that explaining a possibility is still a far cry from explaining a fact: while the differences in the structure of mental maps caused by the details of reindeer herding practices explain, in our opinion, how the further differences in spatial cognition between the groups can exist, they do not explain why they exist. Further research is definitely needed in order to answer this question.

Another example of the "second-order" cognitive specialisations belonging to the second group of cognitive particularities is the different types of microlandscape-based 2-D-S elements on the cognitive maps of Komi and Nenets reindeer herders. As was argued in detail in chapter 5, the difference in the types of these elements between the two groups makes perfect sense in the context of the different cognitive tasks the reindeer herders from these groups face in their everyday practice of reindeer pasturing. The Komi must be able to predict the immediate effects of forage, terrain, and a complex of other factors on reindeer behaviour in order to maneuver the herd. The Nenets, on the other hand, face the task of predicting the cumulative effect of these factors over extensive periods of time in order to locate a dispersed herd. It could be argued that this difference falls short of a "pure" cognitive specialisation because the microlandscape terminologies of Komi and Nenets as collectively shared cultural models (conceptualisation systems) are very likely to have played a role in its formation. This, however, poses another interesting question about the origin of these shared cultural models. Indeed, taking into account the obvious inferential value of the given types of microlandscape-based 2-D-S elements in the context of the specific forms of reindeer herding practice among the two groups, the respective conceptualisation systems could well be the linguistic reflections of the particular ways of representing environmental information spatially that the herders have found useful for solving the cognitive tasks they encounter on a daily basis. In other words, these shared cultural models could originate from the herders' reflections on the cognitive specialisations they have to develop. This view can explain how the conceptualisation of microlandscape among the Komi, despite being very different from that of the Nenets, can still be based on Nenets terms – something that would be rather difficult to explain if one insisted on the viewpoint that the types of 2-D-S elements are defined exclusively by the particular conceptualisation system learned. Furthermore, this view, if it is correct, points to one of the seldom discussed mechanisms of how culturally shaped cognition can have an impact back on culture by modifying the collective representations and models that comprise it. Unfortunately, this interesting topic is beyond the scope of this book, so we shall not pursue it further here.

Finally, the third example of the "second-order" cognitive specialisations is the particular forms of the "theory of reindeer mind" that reindeer herders use to decipher and predict reindeer behaviour. As we argued in chapters 6 and 7, these "theories" are attuned to explain reindeer behaviour, which is itself largely shaped by the specific reindeer herding practice; it can even be supposed that these "theories" represent effective deciphering and predicting tools exactly due to this fact. Indeed, the reasoning of the old Nenets reindeer herder that the animals "wanting to get a break from the mosquitoes will come to the camp by themselves" because they "will reason that the smoke from our fires will drive the mosquitoes away" would hardly find an exact analogue among Komi herders because Komi reindeer never come to the camp by themselves (see chapter 6). The obvious conclusion is that the mental state attributions and inferences from them made by the Nenets herder emerge from his cognitive adaptation to the task of predicting reindeer behaviour shaped by a specific (Nenets) reindeer herding practice. This adaptation would be of little help if the task had been to make sense of and predict the behaviour of Komi reindeer. The difficulties reindeer herders experience in dealing with animals that were transferred from herds of other reindeer herding groups (referred to in chapter 6) attest to this fact. This case points to the second under-researched aspect of cognitive specialisations: in addition to cognitive processes and routines for solving specific cognitive tasks posed by a culture, cognitive specialisations include tools for solving cognitive tasks that involve particular culturally modified elements or occur in specific settings. The tasks themselves, in this case, may be less specific.

As a further illustration of the last point, let us recall the representation of distance between locations in *nedalava*s – practically, in the number of stops one has to make along the way from point A to point B to allow the reindeer to rest – that has been observed among the Nenets herders and shown to play important role in their spatial navigation (chapter 4). It is important to note that this way of representing distance does not seem to be directly related to the reindeer herding practice per se. Indeed, we cannot point out any task in the framework of the Nenets pasturing technology for which it would be of practical advantage. The cognitive tasks for which this representation seems to be used – recognising that one is lost, identification of places, shortcut inference, etc. (see chapter 4) – are by no means specific to a reindeer herding culture.

These tasks are routinely faced by any traveler, regardless of his or her cultural background. Therefore, to the degree to which the representation of distance in *nedalavas* can be treated as a cognitive specialisation to these tasks – and we think that it definitely can be – it falls into the third group of specialisations outlined at the beginning of this section: it is a cognitive specialisation to solve rather general cognitive tasks in the specific settings of a reindeer herding culture. The specific factors that make this cognitive specialisation both possible and efficient in solving the tasks are a) the fact that up until very recently the absolute majority of trips made by reindeer herders were performed on a reindeer sledge, and b) the fact that reindeer have a rather stable pattern of run and rest that herders tend to follow in the course of their trips. As was argued in chapter 4, these two facts made it possible for the herders to include reindeer as an external element in a distributed cognitive system used to solve some of the most essential navigational tasks. Their spatial representations – and particularly the representation of distance – are tailored to work in such a system. In this case, just as in the previous one, culture creates cognitive specialisation by affecting the settings in which cognitive tasks are to be solved.

8.4 Culture, Economy, and Cognition

The empirical material and the analysis presented in this book suggest the following general conclusions about the relationship between culture and a "traditional" economy, on the one hand, and human cognition on the other.

First of all, culture can and does shape human cognition beyond the level of cognitive content. Furthermore, its influence can, in fact, go deeper than the level of conceptualisation and inference schemes that specify the application of certain cognitive processes all the way to the cognitive processes themselves. This means that culture not only specifies the purposes for which and the manner in which specific cognitive tools get applied; it can even engineer the tools themselves, making them more appropriate for the purposes at hand. This, in turn, suggests that members of different cultures can differ in their cognitive abilities related to specific domains.

Of course, differences of this sort cannot result directly from members acquiring their cultural group's shared representations, such as conceptual systems, inference and behavioural scripts, or declarative knowledge in general. Cognitive tools can be modified only through practice – recurrent experience in solving particular practical tasks for which a certain redesign of cognitive tools would be necessary or bring a significant advantage. We believe, and we hope to have convinced the reader, that the power of culture to influence cognitive processes is explained first of all by its ability to shape these activities – and hence the practical experience – of its members. The principal subsistence activity that forms the core of what is usually called a "traditional" economy arguably has pride of place here. This activity is the most mundane, persistent, recurrent, and, at the same time, the most demanding and important in most human societies, and it usually provides the framework for all the other tasks and pursuits in which their members engage. For these reasons, such subsistence activities are particularly well placed to shape cognitive processes. Since both the type and details of these activities are, of course, culturally specific, the traditional economy is likely to be the main channel through which culture influences individual cognition.

The material presented in this book also shows exactly how this cultural influence on cognitive processes can be realised through forms of everyday practice. It seems that by determining the type and details of the principal everyday economic activity of its members, culture also determines the specific cognitive tasks they are likely to encounter on a daily basis, as well as the concrete settings and situations in which these and other less specific cognitive tasks are encountered and have to be solved. The recurrent experience of solving particular cognitive tasks in particular settings leads to the development of cognitive tools that make solving these tasks in these settings most effective. These tools are shared among the members of a culture to the degree that they share the same experiences. We have labeled this process "cognitive specialisation," and demonstrated a number of instances of its work among the members of the two reindeer herding groups in our study. Although we do not argue that cognitive specialisation is the only process that can lead to variation in cognitive tools between members of different cultures, it seems to us that it is among the most important. We have tried to demonstrate in this chapter that most of the specific cognitive traits we have managed to identify among the reindeer herders can be explained from the viewpoint of the cognitive specialisation model.

Finally, the material presented in this book seems to confirm the validity of psychological universalism both as an ontological position and an epistemological approach. While we have asserted that culturally specific cognitive processes do exist, they seem to represent specific variants of universal cognitive processes that can be observed among members of any society or culture. In other words, all humans, irrespective of their cultural, professional, religious, or social identity, share the same basic psychological processes and mechanisms on which culture simply plays variations. In our opinion, this is rather good news from both the scientific and practical points of view. From the scientific point of view, it means that an interpretative study of human behaviour based, for example, on qualitative methods of participant observation is possible and can produce valid results (see the discussion in chapter 1). This is particularly important for a discipline such as cultural anthropology, which is based on qualitative and interpretative methods.

From the practical point of view, this conclusion means that human cultures cannot be so different as to make communication and cooperation between their members impossible. Although the experience of living in a certain culture can equip a human with specific cognitive capacities in specific domains (for example, understanding reindeer behaviour), it cannot make him or her either superhuman or cognitively inferior. As far as cognition is concerned, all humans are indeed born equal, stay equal throughout their lives, and have the possibility of effective cross-cultural communication, cooperation, and peaceful co-existence no matter how different their cultures may be.

LIST OF REFERENCES

Acredolo, L. P. 1981. Small- and Large-Scale Spatial Concepts in Infancy and Childhood. In *Spatial Representation and Behaviour Across the Life Span,* L. Liben, A. Patterson and N. Newcombe (eds.), 63–81. New York: Academic Press.

Adaev, V.N. 2015a. "Metki zemli" i "metki dorogi": nazemnyie prirodnyie i iskusstennyie ob'ekty v sisteme prostranstvennogo orientirovaniia tundrovykh nentsev ['Marks of earth' and 'marks of trails': land-based natural and artificial objects used in the spatial orientation by tundra Nenets]. *Vestnik arkheologii, antropologii i etnografii* 1(28): 98–106.

— 2015b. O mentalnykh kartakh v prostranstvennom orientirovanii nentsev [On mental maps and spatial orientation of the Nenets] . *AB ORIGINE* 7: 94–106.

— 2015c. Formy peredachi geograficheskoi informatsii v traditsii tundrovykh i lesnykh nentsev [Forms of transmission of geographic information among the Tundra and Forest Nenets] (in Russian). *Uralski istoricheski vestnik* 2(47): 33–38.

— 2015d. Orientirovanie po astronomicheskim ob'ektam v traditsionnoi kul'ture nentsev [The traditional Nenets wayfinding with the use of astronomy objects]. In *Chelovek i Sever: Antropologia, arkheologiia, ekologiia,* 210–213. Tiumen: Izdatel'stvo IPOS SO RAN.

— 2016. Nenetskiie priiomy orientirovaniia v puti: ispol'zovaniie pomoshi olenia i sobaki [The Nenets' navigation techniques: using the help of reindeer and dog]. *Vestnik arkheologii, antropologii i etnografii* 1(32): 133–141.

— 2018. Spetsifika orientirovaniia po storonam sveta u tundrovykh nentsev [The Tundra Nenets Ways of Spatial Orientation with theUse of Cardinal Directions] (in Russian). *Vestnik arkheologii, antropologii i etnografii* 1(40): 89–98.

Adaev, V.N. and R.H. Rahimov 2015. Traditsionnaia kul'tura puteshestvii tundrovykh nentsev [Traditional travel culture of the Tundra Nenets]. *Vestnik arkheologii, antropologii i etnografii* 2(29): 151–158.

Adriansen, H. K. and T. T. Nielsen 2002. Going Where the Grass is Greener: On the Study of Pastoral Mobility in Ferlo, Senegal. *Human Ecology* **30**: 215–226.

Aikhenvald, A. Y. 2006. *Evidentiality*. Oxford: Oxford University Press.

Allen, G. L. 1999. Spatial Abilities, Cognitive Maps, and Wayfinding: Bases for Individual Differences in Spatial Cognition and Behaviour. In *Wayfinding Behaviour: Cognitive Mapping and Other Spatial Processes,* R. G. Golledge (ed.), 46–80. Baltimore: The Johns Hopkins University Press.

— 2004. *Human Spatial Memory: Remembering Where.* New York: Lawrence Erlbaum Assoc Inc.

Allwood, C. M. 1998. The Creation and Nature(s) of Indigenized Psychologies from the Perspective of the Anthropology of Knowledge. *Knowledge and Society* 11: 153–172.

Anderson, D. G., J. P. L. Loovers, S. A. Schroer, and R. P. Wishart 2017. Architectures of

Domestication: On Emplacing Human-Animal Relations in the North. *Journal of the Royal Anthropological Institute* 23: 398–416.

Antweiler, C. 1998. Local Knowledge and Local Knowing: An Anthropological Analysis of Contested 'Cultural Products' in the Context of Development. *Anthropos* 93: 469–494.

— 2004. Local Knowledge Theory and Methods: An Urban Model from Indonesia. In *Investigating Local Knowledge: New Directions, New Approaches*, A. Bicker, P. Sillitoe, and J. Pottier (eds.), 1 – 34. Aldershot Hants (England): Ashgate.

Aporta, C. 2009. The Trail as Home: Inuit and Their Pan-Arctic Network of Routes. *Human Ecology* 37: 131–146.

Aporta, C. and E. Higgs 2005. Satellite Culture: Global Positioning Systems, Inuit Wayfinding, and the Need for a New Account of Technology. *Current Anthropology* 46: 729–753.

Appleyard, D. 1970. Styles and Methods of Structuring a City. *Environment and Behavior* 2: 100–117.

Asquith, P. 1984. The Inevitability and Utility of Anthropomorphism in Description of Primate Behaviour. In *The Meaning of Primate Signals*, R. Harre and V. Reynolds (eds.): 138–174. Cambridge: Cambridge University Press.

Atran, S. 1990. *Cognitive Foundations of Natural History: Towards an Anthropology of Science*. Cambridge: Cambridge University Press.

Barrett, J. L., R. M. Newman, and R. A. Richert 2003. When Seeing is not Believing: Children's Understanding of Humans' and Non-Humans' Use of Background Knowledge in Interpreting Visual Displays. *Journal of Cognition and Culture* 3: 91–108.

Barrett, J. L.; R. A. Richert and A. Driesenga 2001. God's Beliefs versus Mother's: The Development of Nonhuman Agent Concepts. *Child Development* 72: 50–65.

Barry, H.; I. L. Child and M. K. Bacon 1959. Relation of Child Training to Subsistence Economy. *American Anthropologist* 61: 51–63.

Barth, F. 1960. The Land Use Patterns of Migratory Tribes of South Persia. *Norsk Geografisk Tidsskrift* 17: 1–11.

— 1961. *Nomads of South-Persia; the Basseri Tribe of the Khamseh Confederacy*. Boston: Little, Brown and Company.

— 2002. An Anthropology of Knowledge. *Current Anthropology* 43: 1–18.

Baskin, L. M. 1968. *Ekologicheskie osnovy severnogo olenevodstva. Avtoreferat na soiskanie uchenoi stepeni kandidata biologicheskikh nauk (in Russian)*. Moscow: Moscow State University.

— 1970. *Severnyi olen': Ekologiia i povedenie [Reindeer: ecology and behaviour]*. Moscow: Nauka.

— 2009. *Severnyi olen': upravlenie povedeniem i populiatsiiami, olenevodstvo, okhota*. Moscow: Tovarishchestvo nauch. izd. KMK.

Bates, D. G. 1972. Differential Access to Pasture in a Nomadic Society: the Yörük of

Southeastern Turkey. In *Perspectives on Nomadism*, W. Irons and N. Dyson-Hudson (eds.), 48–59. Leiden: E. J. Brill.

— 1973. *Nomads and Farmers: A Study of the Yörük of Southeastern Turkey*. (Anthropological Papers of the Museum of Anthropology, University of Michigan 52). Ann Arbor: University of Michigan Press.

Beach, H. 1981. *Reindeer-herd Management in Transition: The Case of Tuorpon Saameby in Northern Sweden*. Uppsala: Upsaliensis Academiae.

— 1990. Comparative Systems of Reindeer Herding. In *The World of Pastoralism: Herding Systems in Comparative Perspective*, J. G. Galaty and D. L. Johnson (eds.), 255–298. London: Belhaven Press.

Beach, H. and F. Stammler 2006. Human-Animal Relations in Pastoralism. *Nomadic Peoples* 10: 6–30.

Belitser, V. N. 1958. *Ocherki po etnografii narodov komi. XIX – nachalo XX vv. [Essays on the ethnography of Komi peoples. 19th – beginning of 20th centuries*. Moscow: Nauka.

Benedict, R. 1934. *Patterns of Culture*. Boston: Houghton Mifflin.

Berkes, F., J. Colding, and C. Folke 2000. Rediscovery of Traditional Ecological Knowledge as Adaptive Management. *Ecological Applications* 10: 1251–1262.

Berlin, B. 1992. *Ethnobiological Classification: Principles of Categorization of Plants and Animals in Traditional Societies*. Princeton: Princeton University Press.

Berry, J. W. 1976a. *Human Ecology and Cognitive Style: Comparative Studies in Cultural and Psychological Adaptation*. London: John Wiley & Sons Inc.

— 1976b. *Human Ecology and Cognitive Style: Comparative Studies in Cultural and Psychological Adaptation*. (Cross-Cultural Research & Methodology Series). New York: Wiley.

— 2004. An Ecological Perspective on the Development of Competence. In *Culture and Competence: Contexts of Life Success*, R. J. Sternberg and E. L. Grigorenko (eds.), 3–22. Washington, DC: American Psychological Association.

Berry, J. W., Y. H. Poortinga, M. H. Segall, and P. R. Dasen 2002. *Cross-Cultural Psychology: Research and Applications*. (2nd edition). Cambridge: Cambridge University Press.

Bock, J. 2005. What Makes a Competent Adult Forager? In *Hunter-Gatherer Childhoods: Evolutionary, Developmental, and Cultural Perspectives*, B. S. Hewlett and M. E. Lamb (eds.), 109–128. New Brunswick N.J.: Aldine Transaction.

De Boer, W. F. and H. H. T. Prins 1989. Decisions of Cattle Herdsmen in Burkina Faso and Optimal Foraging Models. *Human Ecology* 17: 445–464.

Bohbot, V. D., G. Iaria, and M. Petrides 2004. Hippocampal Function and Spatial Memory: Evidence From Functional Neuroimaging in Healthy Participants and Performance of Patients With Medial Temporal Lobe Resections. *Neuropsychology* 18: 418–425.

Borgerhoff Mulder, M. and D. W. Sellen 1994. Pastoralist Decisionmaking: A Behavioral Ecological Perspective. In *African Pastoralist Systems: An Integrated approach,* E. Fratkin, K. A. Galvin, and E. A. Roth (eds.), 205–230. London: Lynne Rienner Publishers.

Botterill, G. 1996. Folk Psychology and Theoretical Status. In *Theories of Theories of Mind*, P. Carruthers and P. K. Smith (eds.), 105–118. Cambridge: Cambridge University Press.

Bourdieu, P. 1977. *Outline of a Theory of Practice.* Cambridge: Cambridge University Press.

Boyd, R. and P. J. Richerson 1985. *Culture and the Evolutionary Process.* Chicago: The University of Chicago Press.

Brunyé, T. T. and H. A. Taylor 2008. Extended Experience Benefits Spatial Mental Model Development With Route But Not Survey Descriptions. *Acta Psychologica* 127: 340–354.

Burnham, P. 1975. 'Regroupment' and Mobile Societies: Two Cameroon Cases. *The Journal of African History* 16: 577–594.

Caporael, L. R. 1986. Anthropomorphism and Mechanomorphism: Two Faces of the Human Machine. *Computers in Human Behavior* 2: 215–234.

Caporael, L. R. and C. M. Heyes 1997. Why Anthropomorphize? Folk Psychology and Other Stories. In *Anthropomorphism, Anecdotes, and Animals,* R. W. Mitchell, N. S. Thompson, and H. L. Miles (eds.), 59–73. New York: State University of New York Press.

Carruthers, P. and P. K. Smith 1996. *Theories of Theories of Mind.* Cambridge: Cambridge University Press.

Casimir, M. J. 1992. The Ecological Determinants of Territorial Behaviour Among Mobile Herders. In *Mobility and Territoriality: Social and Spatial Boundaries Among Foragers, Fishers, Pastoralists, and Peripatelics,* M. J. Casimir and A. Rao (eds.), 153–203. New York: Berg.

Castner, H. W. 1995. Where Am I? Basic Concepts in Human Orientation. *Cartographica: The International Journal for Geographic Information and Geovisualization* 32: 54–56.

Cheney, D. L. and R. M. Seyfarth 1992. *How Monkeys See the World: Inside the Mind of Another Species.* Chicago: University of Chicago Press.

Chernov, Y. I. 1980. *Zhiz'n tundry.* Moscow: Mysl.

Clark, A. 2010. *Supersizing the Mind: Embodiment, Action, and Cognitive Extension.* Oxford: Oxford University Press.

Collett, T. S. and P. Graham 2004. Animal Navigation: Path Integration, Visual Landmarks and Cognitive Maps. *Current Biology* 14: R475–R477.

Coppock, D. L., J. E. Ellis, and D. M. Swift 1986. Livestock Feeding Ecology and Resource Utilization in a Nomadic Pastoral Ecosystem. *The Journal of Applied Ecology* 23: 573–583.

Coppolillo, P. B. 2000. The Landscape Ecology of Pastoral Herding: Spatial Analysis

of Land Use and Livestock Production in East Africa. *Human Ecology* 28: 527–559.
Craik, K. H. 1973. Environmental Psychology. *Annual Review of Psychology* 24: 403–422.
Cronk, L. 1991. Human Behavioral Ecology. *Annual Review of Anthropology* 20: 25–53.
Cronk, L.; N. Chagnon and W. Irons 2000. *Adaptation and Human Behavior: An Anthropological Perspective*. New York: Routledge.
D'Andrade, R. G. 1990. Some Propositions About the Relations Between Culture and Human Cognition. In *Cultural Psychology: Essays on Comparative Human Development,* J. W. Stigler; R. A. Schweder and G. Herdt (eds.), 65–129. Cambridge: Cambridge University Press.
— 1995. *The Development of Cognitive Anthropology*. Cambridge University Press.
D'Andrade, R. G. and C. Strauss 1992. *Human Motives and Cultural Models*. Cambridge: Cambridge University Press.
Davydov, V. N. 2017. Research on Human and Reindeer Relations in Southern Yakutia [Sakha Republic], *Anthropology and Archeology of Eurasia* 56: 6–31.
Delvin, A. S. 1976. The 'Small Town' Cognitive Map: Adjusting to a New Environment. In *Environmental Knowing: Theories, Research and Methods*, G. T. Moore and R. G. Golledge (eds.), 58–66. Stroudsburg: Dowden, Hutchinson and Ross.
Dennett, D. C. 1981. *Brainstorms: Philosophical Essays on Mind and Psychology*. Cambridge Mass.: MIT Press.
— 1987. *The Intentional Stance*. Cambridge Mass.: MIT Press.
— 1995. Do Animals have Beliefs? In *Comparative Approaches to Cognitive Science,* J.-A. Meyer and H. L. Roitblat (eds.), 111–118. Cambridge Mass.: MIT Press.
deWaal, F. 2006. *Primates and Philosophers: How Morality Evolved*. Princeton: Princeton University Press.
Doherty, M. 2009. *Theory of Mind: How Children Understand Others' Thoughts and Feelings*. Hove: Psychology Press.
Dolgikh, B. O. 1970. *Ocherki po etnicheskoi istorii nentsev i entsev*. Moscow: Nauka.
Downs, R. M. and D. Stea 1973. Cognitive Maps and Spatial Behavior: Process and Products. In *Image and Environment: Cognitive Mapping and Spatial Behavior,* R. M. Downs and D. Stea (eds.), 8–26. London: Arnold.
— 1977. *Maps in Minds: Reflections on Cognitive Mapping*. New York: Joanna Cotler Books.
Durkheim, E. 2008. *The Elementary Forms of Religious Life*. (abridged edition). Oxford: Oxford University Press.
Dwyer, M. J. and K. V. Istomin 2006. Mobility and Technology: Understanding the Vulnerability of Two Groups of Nomadic Pastoralists to Reindeer Losses. *Nomadic Peoples* 10: 142–165.
— 2008. Theories of Nomadic Movement: A New Theoretical Approach for Understanding the Movement Decisions of Nenets and Komi Reindeer Herders. *Human Ecology* 36: 521–533.

— 2009. Komi Reindeer Herding: The Effects of Socialist and Post-Socialist Change on Mobility and Land Use. *Polar Research* 28: 282–297.

Dyson-Hudson, N. 1972. The Study of Nomads. *Journal of Asian and African Studies* 7: 2–29.

Dyson-Hudson, R. and N. Dyson-Hudson 1980. Nomadic Pastoralism. *Annual Review of Anthropology* 9: 15–61.

Dyson-Hudson, R. and J.T. McCabe 1985. *South Turkana Nomadism: Coping with an Unpredictably Varying Environment.* (Ethnography Series FL17-001). New Haven: Human Relations Area Files.

Dyson-Hudson, R. and E.A. Smith 1978. Human Territoriality: An Ecological Reassessment. *American Anthropologist* 80: 21–41.

Easton, R.D. and M.J. Sholl 1995. Object-Array Structure, Frames of Reference, and Retrieval of Spatial Knowledge. *Journal of Experimental Psychology: Learning, Memory, and Cognition* 21: 483–500.

Elam, Y. 1979. Nomadism in Ankole as a Substitute for Rebellion. *Africa: Journal of the International African Institute* 49: 147–158.

Ellen, R. 2003. Arbitrariness and Necessity in Ethnobiological Classification: Notes on Some Persisting Issues. In *Nature Knowledge: Ethnoscience, Cognition, and Utility,* G. Sanga and G. Ortalli (eds.), 47–56. New York: Berghahn Books.

— 2006. *The Categorical Impulse: Essays in the Anthropology of Classifying Behaviour.* New York: Berghahn Books.

Enfield, N.J. and S.C. Levinson 2006. *Roots of Human Sociality: Culture, Cognition and Interaction.* Oxford: Berg.

Engels, F. 1907. *Landmarks of Scientific Socialism: Anti-Dühring.* (1st edition). Chicago: C.H. Kerr & Company.

Erdenebaatar, B. 1996. Socio-economic Aspects of the Pastoral Movement Patterns of Mongolian Herders, vol. 1. In *Culture and Environment in Inner Asia,* C. Humphrey and D. Sneath (eds.), 58–110. Cambridge: White Horse Press.

Etienne, A.S.; R. Maurer; J. Georgakopoulos and A. Griffin 1999. Dead Reckoning (Path Integration), Landmarks, and Representation of Space in a Comparative Perspective. In *Wayfinding Behaviour: Cognitive Mapping and Other Spatial Processes,* R.G. Golledge (ed.), 197–228. Baltimore: The Johns Hopkins University Press.

Evans-Pritchard, E.E. 1940. *The Nuer: A Description of the Modes of Livelihood and Political Institutions of a Nilotic People.* Oxford: Clarendon Press.

Feldman, M.W. and K.N. Laland 1996. Gene-Culture Coevolutionary Theory. *Trends in Ecology and Evolution* 11: 453–457.

Foo, P., W.H. Warren, A. Duchon, and M.J. Tarr 2005. Do Humans Integrate Routes Into a Cognitive Map? Map- Versus Landmark-Based Navigation of Novel Shortcuts. *Journal of Experimental Psychology: Learning, Memory, and Cognition* 31: 195–215.

Forde, C. D. 1934. *Habitat, Economy and Society: A Geographical Introduction to Ethnology*. London: Methuen & Co. ltd.

Freeman, D. 1983. *Margaret Mead and Samoa: The Making and Unmaking of an Anthropological Myth*. Harvard University Press.

Galef, B. G. 1993. Functions of Social Learning About Food: A Causal Analysis of Effects of Diet Novelty on Preference Transmission. *Animal Behaviour* 46: 257–265.

Gallistel, C. R. 1993. *The Organization of Learning*. (Reprint). Boston: MIT Press.

Gallup, G. G., L. Marino, and T. J. Eddy 1997. Anthropomorphism and the Evolution of Social Intelligence: A Comparative Approach. In *Anthropomorphism, Anecdotes, and Animals*. R. W. Mitchell, N. S. Thompson, and H. L. Miles (eds.), 77–91. New York: State University of New York Press.

Garling, T., A. Book, E. Lindberg, and T. Nilsson 1981. Memory For the Spatial Layout of the Everyday Physical Environment: Factors Affecting Rate of Acquisition. *Journal of Environmental Psychology* 1: 263–277.

Geertz, C. 1973. *The Interpretation of Cultures: Selected Essays*. New York: Basic Books.

— 1983. *Local Knowledge: Further Essays in Interpretive Anthropology*. New York: Basic Books.

Gell, A. 1985. How to Read a Map: Remarks on the Practical Logic of Navigation. *Man* 20: 271–286.

Gibson, J. J. 1986. *The Ecological Approach to Visual Perception*. London: Lawrence Erlbaum Assoc Inc.

Gladwin, T. 1970. *East Is a Big Bird: Navigation and Logic on Puluwat Atoll*. New York: Harvard University Press.

Golledge, R. 1999. *Wayfinding Behavior: Cognitive Mapping and Other Spatial Processes*. London: Johns Hopkins University Press.

Golledge, R. G. 1999. Human Wayfinding and Cognitive Maps. In *Wayfinding Behaviour: Cognitive Mapping and Other Spatial Processes*, R. G. Golledge (ed.), 5–45. London: Johns Hopkins University Press.

Golledge, R. G., G. D. Richardson, J. N. Rayner, and J. J. Parnicky 1983. Procedures for Defining and Analyzing Cognitive Maps of the Mildly and Moderately Mentally Retarded. In *Spatial Orientation: Theory, Research, and Application*, H. L. Pick and Linda P. Acredolo (eds.), 79–104. New York: Plenum Press.

Golledge, R. G. and A. N. Spector 1978. Comprehending Urban-Environment: Theory and Practice. *Geographical Analysis* 10: 403–426.

Golovnev, A. V. 1995. *Govoriashchie kul'tury: tradicii samodijcev i ugrov*. Ekaterinburg: Institut istorii i arckheologii UrO RAN.

Golovnev, A. V. and G. Osherenko 1999. *Siberian Survival: The Nenets and Their Story*. Ithaca NY: Cornell University Press.

Golovnev, A. V. 2009. Comment on 'Finding the Way': A Critical Discussion of Anthropological Theories of Human Spatial Orientation with Reference to Reindeer Herders of North-Eastern Europe and Western Siberia, by Kirill V. Istomin and Mark J. Dwyer. *Current Anthropology* 50: 43.

Goodenough, W.H. 1957. Cultural Anthropology and Linguistics. In *Report of the Seventh Annual Round Table Meeting on Linguistics and Language Study,* P.L. Garvin (ed.), 167–177. Washington: Georgetown University Press.

Gopnik, A. and H. M. Wellman 1992. Why the Child Theory of Mind Really is a Theory. *Mind and Language* 7: 145–171.

Gordon, R.M. 1986. Folk Psychology as Simulation. *Mind and Language* 1: 158–171.

Gulliver, P.H. 1975. Nomadic Movements: Causes and Implications. In *Pastoralism in Tropical Africa, Th.* Monod (ed.), 369–386. Oxford: Oxford University Press.

Habeck, J. O. 2005. *What It Means to be a Herdsman: The Practice and Image of Reindeer Husbandry Among the Komi of Northern Russia.* Münster: LIT.

Harris, P.L. 1992. From Simulation to Folk Psychology: The Case for Development. *Mind and Language* 7, 120–144.

— 1996. Desires, Beliefs, and Language. In *Theories of Theories of Mind,* P. Carruthers and P. K. Smith (eds.), 200–222. Cambridge: Cambridge University Press.

Hart, H. and A. Pantzer 1925. Have Subhuman Animals Culture? *The American Journal of Sociology* 30: 703–709.

Hart, R.A. and G. T. Moore 1973. The Development of Spatial Cognition: A Review. In *Image and Environment: Cognitive Mapping and Spatial Behaviour,* R.M. Downs and D. Stea (eds.), 246–288. London: Arnold.

Haviland, J. 2000. Pointing, Gesture Spaces, and Mental Maps. In *Language and Gesture,* D. McNeill (ed.), 13–46. Cambridge: Cambridge University Press.

Heal, J. 1986. Replication and Functionalism. In *Language, Mind and Logic,* J. Butterfield (ed.), 135–150. Cambridge: Cambridge University Press.

Heft, H. 1996. The Ecological Approach to Navigation: A Gibsonian Perspective. In *The Construction of Cognitive Maps,* J. Portugali (ed.), 105–132. Dordrecht: Kluwer Academic Publishers.

Herskovits, M.J. 1948. *Man and His Works: The Science of Cultural Anthropology.* (First edition). New York: Alfred A Knopf.

Heyes, C.M. 1994. Social Learning in Animals – Categories and Mechanisms. *Biological Reviews of the Cambridge Philosophical Society* 69: 207–231.

Heyes, C.M. and B.G. Galef 1996. *Social Learning in Animals: The Roots of Culture.* San Diego: Academic Press.

Hofstede, G. H. 1980. *Culture's Consequences: International Differences in Work-Related Values.* (Expanded). Beverly Hills: Sage Publ. Inc.

Holland, D. and N. Quinn 1987. *Cultural Models in Language and Thought.* Cambridge: Cambridge University Press.

Holmes, L.D. 1988. *Quest for the Real Samoa: The Mead/Freeman Controversy and Beyond.* (New edition). South Hadley: Bergin & Garvey Publ. Inc..

Hunn, E.S. 1977. *Tzeltal Folk Zoology: The Classification of Discontinuities in Nature.* New York: Academic Press.

Huntington, H. P. 2000. Using Traditional Ecological Knowledge in Science: Methods and Applications. *Ecological Applications* 10: 1270–1274.
Hutchins, E. 1995. *Cognition in the Wild.* Cambridge Mass.: MIT Press.
— 2010. Cognitive Ecology. *Topics in Cognitive Science* 2, 705–715.
Ingold, T. 1974. On Reindeer and Men. *Man* 9: 523–538.
— 1986. *The Appropriation of Nature: Essays on Human Ecology and Social Relations.* Manchester: Manchester University Press.
— 1988. *Hunters, Pastoralists and Ranchers: Reindeer Economies and their Transformations.* (Cambridge Studies in Social and Cultural Anthropology). Cambridge: Cambridge University Press.
— 1990. An Anthropologist Looks at Biology. *Man* 25: 208–229.
— 1996. The Optimal Forager and Economic Man. In *Nature and Society: Anthropological perspectives,* P. Descola and G. Pálsson (eds.), 25–44. London: Routledge.
— 2000. *The Perception of the Environment: Essays on Livelihood, Dwelling and Skill.* London: Routledge.
— 2003. Two Reflections on Ecological Knowledge. In *Nature Knowledge: Ethnoscience, Cognition, and Utility,* G. Sanga and G. Ortalli (eds.), 301–311. New York: Berghahn Books.
Irons, W. 1974. Nomadism as a Political Adaptation: The Case of the Yomut Turkmen. *American Ethnologist* 1: 635–658.
Ishikawa, T. and D. R. Montello 2006. Spatial Knowledge Acquisition from Direct Experience in the Environment: Individual Differences in the Development of Metric Knowledge and the Integration of Separately Learned Places. *Cognitive Psychology* 52, 93–129.
Islavin, V. 1847. *Samoedy v domashnem i obshchestvennom bytu [Samojeds in their private and social life].* St. Petersburg: Ministerstvo gosudarstvennykh imushestv.
Istomin, K. V. 2000. Living in Chum: Social Relations and Personal Behavioural Strategies among Komi Reindeer Herders. *Pro Ethnologia: Publications of the Estonian National Museum* 10, 49–81.
— 2004. *Etnoekologicheskaia kharakteristika komi-izhemskogo olenevodstva* [Ethnoecological characteristics of the Iyhma Komi reindeer herding]. Ph.D. Thesis. Syktyvkar-Moscow Russian Academy of Science.
— 2005. Komi-Izhemskoe olenevodstvo. Istoriografiia dvukh vekov izucheniia. *Studia Juvenalia* (2): 88–102.
— 2010. Komi-izhemskoe olenevodstvo Bolshezemelskoi tundry [Izhma-komi reindeer herding of Bolshezemelskaia tundra]. In *Pripolarnaia perepis 1926/27 gg. na Evropeiskom Severe (Arkhangelskaia gubernia i avtonomnaia oblast' Komi),* K. B. Klokov and D. P. Ziker (eds.), 317–339. St. Petersburg: MPSS.
— 2011. The Land to Herd and the Space to Travel: Comparing the Categorizations of Landscape Among Komi and Nenets Reindeer Herding Nomads. In *Nomadismus in der 'Alten Welt': Formen der Repräsentation in Vergangenheit und Gegenwart,* (L.

Prager (Hg.), 233–256. Münster: LIT Verlag.
— 2017. O dinamike kul'tury olenei na Kolskom poluostrove [The dynamics of reindeer culture on the Kola Peninsula]. *Uralski istoricheski vestnik* 2(55): 16–24.
Istomin, K.V. and M.J. Dwyer 2009. Finding the Way: A Critical Discussion of Anthropological Theories of Human Spatial Orientation with Reference to Reindeer Herders of Northeastern Europe and Western Siberia. *Current Anthropology* 50: 29–49.
— 2010. Dynamic Mutual Adaptation: Human-Animal Interaction in Reindeer Herding Pastoralism. *Human Ecology* 38: 613–623.
Jamieson, D. 2009. What Do AnimalsThink? In *The Philosophy of Animal Minds,* R.W. Lutz (ed.), 15–34. Cambridge: Cambridge University Press.
Jernsletten, J.-L.L. and K.B. Klokov 2002. *Sustainable Reindeer Husbandary.* Tromsø: University of Tromsø.
Johnson, S.C. 2003. Detecting Agents. *Philosophical Transactions: Biological Sciences* 358: 549–559.
Kaplan, S. 1973. Cognitive Maps in Perception and Thought. In *Image and Environment: Cognitive Mapping and Spatial Behaviour,* R.M. Downs and D. Stea (eds.), 63–78. London: Arnold.
— 1976. Adaptation, Structure and Knowledge. In *Environmental Knowing: Theories, Research and Methods,* G.T. Moore and R.G. Golledge (eds.), 32–45. Stroudsburg, PA: Dowden, Hutchinson and Ross.
Kennedy, J.S. 1992. *The New Anthropomorphism.* Cambridge: Cambridge University Press.
Kertselli, S.V. 1911. *Po Bol'shezemel'skoi Tundre s kochevnikami [Through the Bolshezemelskaia Tundra with nomads].* Archangelsk: Gubernskaia tipografiia.
Khomich, L.V. 1995. *Nentsy. Ocherki traditsionnoi kul'tury [Nenets. Essays on traditional culture].* St. Petersburg: Russki dvor.
Kim, U., K.-S. Yang, and K.-K. Hwang 2006. *Indigenous and Cultural Psychology: Understanding People in Context.* New York:Springer.
Kitayama, S., S. Duffy, and Y. Uchida 2007. Self as Cultural Mode of Being. In *Handbook of Cultural Psychology,* S. Kitayama and D. Cohen (eds.), 136–174. New York: The Guilford Press.
Kitchin, R.M. 1994. Cognitive Maps: What Are They and Why Study Them? *Journal of Environmental Psychology* 14: 1–19.
— 1996. Are There Sex Differences in Geographic Knowledge and Understanding? *The Geographical Journal* 162: 273–286.
Knight, N. 2008. Yukatek Maya Children's Attributions of Belief to Natural and Non-natural Entities. *Journal of Cognition and Culture* 8: 235–243.
Konakov, N.D. 1983. *Komi okhotniki i rybolovy vo vtoroi polovine XIX – nachale XX vekov: kul'tura promyslovogo naseleniia taezhnoi zony Evropeiskogo Severo-Vostoka [Komi hunters and fishermen in the second half of the 19th – beginning of the 20th*

centuries: the culture of hunting inhabitants of the taiga zone in the European North-East]. Moscow: Nauka.
Konakov, N.D. and O.V. Kotov 1991. *Etnoarealnye gruppy komi: formirovanie i sovremennoe etnokulturnoe sostoianie*. Moscow: Nauka.
Krupnik, I. 1993. *Arctic Adaptations: Native Whalers and Reindeer Herders of Northern Eurasia*. Hanover: Dartmouth College.
Krupnik, I.I. 1976. Stanovlenie krupnotabunnogo olenevodstva u tundrovykh nentsev. *Sovetskaia etnografiia* (2): 57–69.
Kuhn, T.S. 1962. *The Structure of Scientific Revolutions*. University of Chicago Press.
Kulhavy, R.W. and W.A. Stock 1996. How Cognitive Maps are Learned and Remembered. *Annals of the Association of American Geographers* 86: 123–145.
Laland, K.N. 1996. Is Social Learning Always Locally Adaptive? *Animal Behaviour* 52: 637–640.
Laland, K.N. and W. Hoppitt 2003. Do Animals Have Culture? *Evolutionary Anthropology: Issues, News, and Reviews* 12: 150–159.
Laland, K.N., P.J. Richerson, and R. Boyd 1996. Developing a Theory of Animal Social Learning. In *Social Learning in Animals: The Roots of Culture,* C.M. Heyes and B. G. Galef (eds.), 129–154. San Diego: Academic Press.
Lashuk, L. P. 1958. *Ocherk etnicheskoi istorii Pechorskogo Kraia*. Syktyvkar: Komi Knizhnoe Izdatel'stvo.
Lattimore, O. 1940. *Inner Asian Frontiers of China*. New York: American Geographical Association.
Lave, J. 1988. *Cognition in Practice: Mind, Mathematics, and Culture in Everyday Life*. Cambridge: Cambridge University Press.
— 1996. The Savagery of Domestic Mind. In *Naked Science: Anthropological Inquiry Into Boundaries, Power, and Knowledge,* L. Nader (ed.), 87–100. New York: Routledge.
Lawton, C.A. and J. Kallai 2002. Gender Differences in Wayfinding Strategies and Anxiety About Wayfinding: A Cross-Cultural Comparison. *Sex Roles* 47: 389–401.
Lefebvre, L. and L.-A. Giraldeau 1996. Is Social Learning an Adaptive Specialization? In *Social Learning in Animals: The Roots of Culture,* C.M. Heyes and B.G. Galef (eds.), 107–128. San Diego: Academic Press.
Lehtisalo, T. 1956. Reiseberichtungen. In *Juraksamojedisches Wörterbuch*, VII–CII. Helsinki: Suomalais-ugrilainen Seura.
Leiser, D. and A. Zilbershatz 1989. The Traveller: A Computational Model of Spatial Network Learning. *Environment and Behavior* 21: 435–463.
Leslie, A.M. 1987. Pretense and Representation: The Origins of 'Theory of Mind'. *Psychological Review* 94: 412–426.
— 1996. A Theory of Agency. In *Causal Cognition: A Multidisciplinary Debate,* D. Sperber; D. Premack and A.J. Premack (eds.), 121–149. Oxford: Oxford University Press.

Levine, M. 1982. You-Are-Here Maps: Psychological Considerations. *Environment and Behavior* 14: 221–237.

Levinson, S.C. 1996. Language and Space. *Annual Review of Anthropology* 25: 353–382.

— 2003. *Space in Language and Cognition: Explorations in Cognitive Diversity*. Cambridge: Cambridge University Press.

Levi-Strauss, C. 1966. *The Savage Mind*. University of Chicago Press.

Lewis, D. 1972. *We, The Navigators: The Ancient Art of Landfinding in the Pacific*. Canberra: Australian National University Press.

— 1976. Observations on Route Finding and Spatial Orientation Among the Aboriginal Peoples of the Western Desert Region of Central Australia. *Oceania* 46, 249–282.

Liebenberg, L. 2001. *The Art of Tracking: The Origin of Science*. Claremont: David Philip.

Little, M.A. and P.W. Leslie 1999. *Turkana Herders of the Dry Savana: Ecology and Biobehavioural Response of Nomads to an Uncertain Environment*. Oxford: Oxford University Press.

Loomis, J.M., R.L. Klatzky, R.G. Golledge, et al. 1993. Nonvisual Navigation by Blind and Sighted: Assessment of Path Integration Ability. *Journal of Experimental Psychology: General* 122: 73–91.

Loomis, J.M., R.L. Klatzky, R.G. Golledge, and J.W. Philbeck 1999. Human Navigation by Path Integration. In *Wayfinding Behavior: Cognitive Mapping and Other Spatial Processes,* R.G. Golledge (ed.): 125–151. London: Johns Hopkins University Press.

Lutz, R.W. 2009. The Philosophy of Animal Minds: An Introduction. In *The Philosophy of Animal Minds,* R.W. Lutz (ed.), 1–14. Cambridge: Cambridge University Press.

Lynch, K. 1960. *The Image of the City*. Cambridge Mass.: M.I.T. Press.

MacEachren, A.M. 1992. Application of Environmental Learning Theory to Spatial Knowledge Acquisition from Maps. *Annals of the Association of American Geographers* 82: 245–274.

Maguire, E. A., K. Woollett, and H. J. Spiers 2006. London Taxi Drivers and Bus Drivers: A Structural MRI and Neuropsychological Analysis. *Hippocampus* 16: 1091–1101.

Maksimov, S. V. 1984. *God na severe [One Year in the North]*. Archangelsk: Severozapadnoe izdatel'stvo.

Malinowski, B. 1939. The Group and the Individual in Functional Analysis. *American Journal of Sociology* 44: 938–962.

Marcus, G.E. and M.M. Fischer 1983. *Anthropology as Cultural Critique: An Experimental Moment in the Human Sciences*. (2nd edition). University of Chicago Press.

Markus, H.R. and M.G. Hamedani 2007. Sociocultural Psychology: The Dynamic Interdependence among Self Systems and Social Systems. In *Handbook of Cultural*

Psychology, S. Kitayama and D. Cohen (eds.), 3–39. New York: The Guilford Press.

Markus, H. R. and S. Kitayama 1991. Culture and the Self: Implications for Cognition, Emotion, and Motivation. *Psychological Review* 98: 224–253.

McCabe, J. T. 1994. Mobility and Land Use Among African Pastoralists: Old Conceptual Problems and New Interpretations. In *African Pastoral Systems: An Integrated Approach*, E. Fratkin; K. A. Galvin and E. A. Roth (eds.), 69–90. London: Lynne Rienner Publishers.

McDonald, K. 2007. Cross-Cultural Comparison of Learning in Human Hunting: Implications for Life-History Evolution. *Human Nature* 18: 386–402.

McNamara, T. P., J. K. Hardy, and S. C. Hirtle 1989. Subjective Hierarchies in Spatial Memory. *Journal of Experimental Psychology: Learning, Memory, and Cognition* 15: 211–227.

Mead, M. 1928. *Comming of Age in Samoa: A Psychological Study of Primitive Youth for Western Civilisation*. New York: William Morrow & Company.

— 1930. *Growing up in New Guinea: A Comparative Study of Primitive Education*. New York: Blue Ribbon Books.

Michrina, B. P. and C. Richards 1996. *Person to Person: Fieldwork, Dialogue, and the Hermeneutic Method*. New York: State University of New York Press.

Miller, G. F. 1941. *Istoriia Sibiri [History of Siberia]*, vol. 2. Moskva: Izdatel'stvo akademii nauk.

Mitchell, R., N. S. Thompson, and H. L. Miles 1997. *Anthropomorphism, Anecdotes, and Animals*. Albany: State University of New York Press.

Mitchell, R. W. 1997. Anthropomorphic Anecdotalism As Method. In *Anthropomorphism, Anecdotes, and Animals*, R. W. Mitchell; N. S. Thompson and H. L. Miles (eds.), 151–169. Albany: State University of New York Press.

Mitchell, R. W. and M. Hamm 1997. The Interpretation of Animal Psychology: Anthropomorphism or Behavior Reading? *Behaviour* 134: 173–204.

Montello, D. 1991. The Measurement of Cognitive Distance: Methods and Construct validity. *Journal of Environmental Psychology* 11: 101–122.

Montello, D. R., K. L. Lovelace, R. G. Golledge, and C. M. Self 1999. Sex-related Differences and Similarities in Geographic and Environmental Spatial Abilities. *Annals of the Association of American Geographers* 89: 515–534.

Moore, G. T. 1976. Theory and Research on the Development of Environmental Knowing. In *Environmental Knowing: Theories Research and Methods*, G. T. Moore and R. G. Golledge (eds.), 138–164. Stroudsburg, PA: Dowden, Hutchinson and Ross.

Moore, G. T. and R. G. Golledge 1976. *Environmental Knowing: Theories, Research and Methods*. Stroudsburg, PA: Dowden, Hutchinson and Ross.

Moore, H. L. and T. Sanders 2006. *Anthropology in Theory: Issues in Epistemology*. Malden MA: Blackwell Pub.

Myres, J. L. 1941. Nomadism. *The Journal of the Royal Anthropological Institute of Great Britain and Ireland* 71: 19–42.

Nelson, R. K. 1969. *Hunters of the Northern Ice*. Chicago: University of Chicago Press.

Newcombe, N. S. 1997. New Perspectives on Spatial Representation: What Different Tasks Tell Us About How People Remember Locations? In *Spatial Cognition in the Child and Adult*, N. Foreman and R. Gillett (eds.), 85–102. (Handbook of Spatial Research Paradigms and Methodologies 1). Hove East Sussex (England): Psychology Press.

Newcombe, N. S. and J. Huttenlocher 2003. *Making Space: The Development of Spatial Representation and Reasoning*. London: MIT Press.

Nichols, S. and S. P. Stich 2003. *Mindreading: An Integrated Account of Pretence, Self-Awareness, and Understanding Other Minds*. Oxford: Oxford University Press.

Nisbett, R. E. 2003. *The Geography of Thought: How Asians and Westerners Think Differently – and why*. New York: Free Press.

Nisbett, R. E., K. Peng, I. Choi, and A. Norenzayan 2001. Culture and Systems of Thought: Holistic Versus Analytic Cognition. *Psychological Review* 108: 291–310.

Nygren, A. 1999. Local Knowledge in the Environment – Development Discourse: From Dichotomies to Situated Knowledge. *Critique of Anthropology* 19: 267–288.

Oehler, A. C. 2020. *Beyond Wild and Tame: Soiot Encounters in a Sentient Landscape*. New Haven and London: Berghahn Books.

O'Keefe, J. and L. Nadel 1978. *The Hippocampus as a Cognitive Map*. Oxford: Oxford University Press.

Orlove, B. S. 1980. Ecological Anthropology. *Annual Review of Anthropology* 9: 235–273.

Paine, R. 1972. The Herd Management of Lapp Reindeer Pastoralists. In *Perspectives on Nomadism*, W. Irons and N. Dyson-Hudson (eds.), 76–87. Leiden: E. J. Brill.

— 1994. *Herds of the Tundra: A Portrait of Saami Reindeer Pastoralism*. Washington: Smithsonian Institution Press.

Passini, R. 1980. Wayfinding – A Conceptual Framework. *Man-Environment Systems.* 10: 22–30.

Perner, J. 1991. *Understanding the Representational Mind.* Cambridge Mass.: MIT Press.

— 1996. Simulation as Explification of Predication-Implicit Knowledge About the Mind: Argumants For a Simulation-Theory Mix. In *Theories of Theories of Mind*, P. Carruthers and P. K. Smith (eds.), 90–104. Cambridge: Cambridge University Press.

Perner, J. and A. Kühberger 2006. Framing and the Theory-Simulation Controversy: Predicting People's Decisions. In *Biological and Cultural Bases of Human Inference*, R. Viale, D. Andler, and L. Hirschfeld (eds.), 161–176. Mahwah, N.J.: Lawrence Erlbaum Associates, Inc.

Piaget, J. and B. Inhelder 1967. *The Child's Conception of Space*. New York: W. W. Norton.

Plato 2002. *Phaedrus*. Boston: IndyPublish.

Pomishin, S. B. 1981. *Problema porody i ieio sovershenstvovanie v olenevodstve [The problem of reindeer races and their development]*. Yakutsk: Yakutskoe knizhnoe izdatel'stvo.

— 1990. *Proiskhozhdenie olenevodstva i domestikatsiia severnogo olenia*. Moscow: Nauka.

Popper, K. R. 1992. *The Logic of Scientific Discovery*. London: Routledge.

Porteus, S. D. 1937. *Primitive Intelligence and Environment*. New York: Macmillan.

Povinelli, D. J. 1997. Panmorphism. In *Anthropomorphism, Anecdotes, and Animals*, R. W. Mitchell, N. S. Thompson, and H. L. Miles (eds.), 92–103. New York: State University of New York Press.

Prestopnik, J. L. and B. Roskos-Ewoldsen 2000. The Relations Among Wayfinding Strategy Use, Sense of Direction, Sex, Familiarity, and Wayfinding Ability. *Journal of Environmental Psychology* 20: 177–191.

Pushkareva, E. T. 2003. *Istoricheskaia tipologiia i etnicheskaia specifika nenetskikh mifov-skazok*. Moscow: Mysl.

Quinn, N. and D. Holland 1987. Culture and Cognition. In *Cultural Models in Language and Thought*, N. Quinn and D. Holland (eds.), 3–40. Cambridge: Cambridge University Press.

Rizzolatti, G. and C. Sinigaglia 2008. *Mirrors in the Brain: How Our Minds Share Actions, Emotions, and Experience*. Oxford: Oxford University Press.

Robbins, P. 1998. Nomadization in Rajasthan, India: Migration, Institutions, and Economy. *Human Ecology* 26: 87–112.

Røed, K. H., Ø. Flagstad, M. Nieminen, et al. 2008. Genetic Analyses Reveal Independent Domestication Origins of Eurasian Reindeer. *Proceedings of the Royal Society B: Biological Sciences* **275**: 1849–1855.

Rupert, R. D. 2009. *Cognitive Systems and the Extended Mind*. Oxford: Oxford University Press.

Sapolsky, R. M. 2006. Social Cultures Among Nonhuman Primates. *Current Anthropology* 47: 641–656.

Savelyeva, E. A. 1971. *Perm Vychegodskaia: K voprosu o proiskhozhdenii naroda komi [Perm of Vzchegda: on the problem of the origin of Komi people]*. Moscow: Nauka.

Schank, R. C. and R. P. Abelson 1977. *Scripts, Plans, Goals and Understanding: Inquiry into Human Knowledge Structures*. John Wiley & Sons Inc.

Schareika, N. 2001. Environmental Knowledge and Pastoral Migration among the Wodaabe of South-Eastern Niger. *Nomadic Peoples* 5: 65–88.

Schrenck, A. G. 1848. *Reise nach dem Nordosten europäischen Russlands durch die Tundren der Samojeden zum Arktischen Uralgebirge*. Dorpat (Tartu): Heinrich Lautmann.

Semple, E. C. 1933. *Influences of Geographic Environment*. London: Constable & Co. Ltd.

Shemyakin, F. N. 1962. General Problems of Orientation in Space and Space Represen-

tations, vol. 1. In *Psychological science in the U.S.S.R.*, B. G. Anan'ev, (ed.), 186–255. Washington: Office of Technical Services.
Shore, B. 1998. *Culture in Mind: Cognition, Culture and the Problem of Meaning.* (New Ed.). New York: Oxford University Press.
Shweder, R. A. 1991. *Thinking Through Cultures: Expeditions in Cultural Psychology.* New York: Harvard University Press.
Sieff, D. F. 1997. Herding Strategies of the Datoga Pastoralists of Tanzania: Is Household Labor a Limiting Factor? *Human Ecology* 25: 519–544.
Siegel, A. W. and S. H. White 1975. The Development of Spatial Representations of Large-Scale Environments. In *Advances in Child Development and Behavior 10,* H. W. Reese (ed.), 9–55. New York: Academic Press.
Silverman, I. and M. Eals 1992. Sex Differences in Spatial Abilities: Evolutionary Theory and Data. In *The Adapted Mind: Evolutionary Psychology and the Generation of Culture,* J. H. Barkow, L. Cosmides, and J. Tooby (eds.), 533–553. Oxford: Oxford University Press.
Simon, H. A. 1957. *Models of Man: Social and Rational.* New York: Wiley.
Skogland, T. 1989. Comparative Social Organization of Wild Reindeer in Relation to Food, Mates and Predator Avoidance. *Advances in Ethology* 29: 1–74.
— 1991. What Are the Effects of Predators on Large Ungulate Populations? *Oikos* 61: 401–411.
Skolkin, V. L. 2008. *Itogi vserossiiskoi selskohozaistvennoi perepisi 2006 goda v 9 tomakh. Tom 7: selskoe khozaistvo raionov krainego severa i priravnennykh k nim mestnostei. [Results of the Russian national agricultural census of 2006 in 9 volumes. Volume 7: agriculture in the regions of far north and the regions that are equal to them in status.* Moscow: Statistika Rossii.
Slezikine, Y. 1996. *Arctic Mirrors: Russia and the Small Peoples of the North.* (New edition). London: Cornell University Press.
Smith, E. A. 1983. Anthropological Applications of Optimal Foraging Theory: A Critical Review. *Current Anthropology* 24: 625–651.
— 2000. Three styles in the Evolutionary Analysis of Human Behaviour. In *Adaptation and Human Behavior: An Anthropological Perspective,* L. Cronk, N. Chagnon, and W. Irons (eds.), 27–46. New York: Aldine.
Spencer, C. and M. Weetman 1981. The Microgenesis of Cognitive Maps: A Longitudinal of New Residents of an Urban Area. *Transactions of the Institute of British Geographers* 6: 375–384.
Spiers, H. J. and E. A. Maguire 2006. Thoughts, Behaviour, and Brain Dynamics During Navigation in the Real World. *NeuroImage* 31: 1826–1840.
— 2008. The Dynamic Nature of Cognition During Wayfinding. *Journal of Environmental Psychology* 28: 232–249.
Spodina, V. I. 2001. *Predstavlenie o prostranstve v traditsionnom mirovozzrenii lesnykh nentsev Nizhnevartovskogo rayona [The concept of space in the worldview of forest*

Nenets from the Niyhnevartovsk region]. Novosibirsk: Agro.

Spooner, B. 1972. The Status of Nomadism as a Cultural Phenomenon in the Middle East. In *Perspectives on Nomadism,* W. Irons and N. Dyson-Hudson (eds.), 122–131. Leiden: E. J. Brill.

Stammler, F. 2005. *Reindeer Nomads Meet the Market: Culture, Property and Globalisation at the 'End of the land'*. (Halle Studies in the Anthropology of Eurasia: HSAE / Max Planck Institute for Social Anthropology). Münster: LIT.

Startsev, G. 1926. Vliianie samoedov na izhemskikh zyrian (iazyk i byt). *Komi mu* 31–34.

Stea, D. and J. M. Blaut 1973. Some Preliminary Observations on Spatial Learning in School Children. In *Image and Environment: Cognitive Mapping and Spatial Behaviour,* R. M. Downs and D. Stea (eds.), 226–234. London: Arnold.

Stépanoff, C. 2012. Human-Animal 'Joint Commitment' in a Reindeer Herding System. *HAU: Journal of Ethnographic Theory* 2: 287–312.

— 2017. The Rise of Reindeer Pastoralism in Northern Eurasia: Human and Animal Motivations Entangled. *Journal of the Royal Anthropological Institute* 23: 376–397.

Stépanoff, C., C. Marchina, C. Fossier and N. Bureau 2017. Animal Autonomy and Intermittent Coexistences: North Asian Modes of Herding. *Current Anthropology* 58: 57–81.

Stépanoff, C. and J. D. Vigne 2018. Hybrid Communities: Biosocial Approaches to Domestication and Other Trans-Species Relationships. New York: Routledge.

Stenning, D. J. 1959. *Savannah Nomads: A Study of the Woodaabe Pastoral Fulani of Western Bornu Province Northern Region, Nigeria*. London: Oxford University Press.

Stephen A. T. 1969. *Cognitive Anthropology*. (illustrated edition). New York: Holt McDougal.

Stevens, A. and P. Coupe 1978. Distortions in Judged Spatial Relations. *Cognitive Psychology* 10: 422–437.

Steward, J. H. 1955. *Theory of Culture Change: The Methodology of Multilinear Evolution*. Urbana: University of Illinois Press.

Strauss, C. and N. Quinn 1998. *A Cognitive Theory of Cultural Meaning*. New York: Cambridge University Press.

Swidler, W. W. 1972. Some Demographic Factors Regulating the Formation of Flocks and Camps Among the Brahui of Baluchistan. In *Perspectives on Nomadism,* W. Irons and N. Dyson-Hudson (eds.), 69–75. Leiden: E. J. Brill.

Tetzlaff, M. and G. Rey 2009. Systematicity and Intentional Realism in Honeybee Navigation. In *The Philosophy of Animal Minds,* R. W. Lutz (ed.), 72–88. Cambridge: Cambridge University Press.

Thomas, E. M. 1993. *The Hidden Life Of Dogs*. Boston: Houghton Mifflin.

Tolman, E. C. 1948. Cognitive Maps in Mice and Men. *The Psychological Review* 55, 189–208.

Tomasello, M. 2001. *The Cultural Origins of Human Cognition*. Boston: Harvard University Press.

Tremlin, T. 2006. *Minds and Gods: The Cognitive Foundations of Religion*. New York: Oxford University Press.

Turner, M. D. 1999. Labor Process and the Environment: The Effects of Labor Availability and Compensation on the Quality of Herding in the Sahel. *Human Ecology* 27: 267–296.

Vasilevich, G. M. and M. G. Levin 1951. Tipy olenevodstva i ikh proiskhozhdenie [Types of reindeer herding and their origin]. *Sovetskaia etnografiia* 1: 63–87.

Vasilyeva, V. V. 2018. Orientirovaniie v prostranstve u dolgan taimyrskogo poberezhiia Severnogo Ledovitogo Okeana [Orientation in Space among the Dolgans of the Taimyr Coast of the Arctic Ocean] . *Uralski istoricheski vestnik* 3(60): 31–39.

Vijver, F. J. R. van de, A. Chasiotis, and S. M. Breugelmans 2011. *Fundamental Questions in Cross-Cultural Psychology*. (1st edition). Cambridge: Cambridge University Press.

Vladimirova, V. 2006. *Just Labor: Labor Ethic in a Post-Soviet Reindeer Herding Community.* (Uppsala Studies in Cultural anthropology / Universitas Upsaliensis. Uppsala: University Press.

Vygotsky, L. S. 1978. *Mind in Society: Development of Higher Psychological Processes*. Harvard: Harvard University Press.

Wang, R. F. 2012. Theories of Spatial Representations and Reference Frames: What Can Configuration Errors Tell Us? *Psychonomic Bulletin and Review* 19: 575–587.

Wang, R. F. and E. S. Spelke 2002. Human Spatial Representation: Insights From Animals. *Trends in Cognitive Sciences* 6: 376–382.

Wassmann, J. and P. R. Dasen 1998. Balinese Spatial Orientation: Some Empirical Evidence of Moderate Linguistic Relativity. *Journal of the Royal Anthropological Institute* 4: 689–711.

Wenger, E. 1999. *Communities of Practice: Learning, Meaning, and Identity*. New York: Cambridge University Press.

Wenger, E. and J. Lave 2006. *L'apprendimento situato. Dall'osservazione alla partecipazione attiva nei contesti sociali*. Centro Studi Erickson.

Whiten, A. 1996. When Does Smart Behaviour-Reading Become Mind-Reading? In *Theories of Theories of Mind,* P. Carruthers and P. K. Smith (eds.), 277–292. Cambridge: Cambridge University Press.

Whiting, B., J. W. M. Whiting, and R. Longabaugh 1979. *Children of Six Cultures: A Psycho-cultural Analysis*. (New edition). Boston: Harvard University Press.

Widlok, T. 1997. Orientation in the Wild: The Shared Cognition of Hai||om bushpeople. *Journal of the Royal Anthropological Institute* 3: 317–332.

Willerslev, R. 2007. *Soul Hunters: Hunting, Animism, and Personhood Among the Siberian Yukaghirs*. Ann Arbor: Univ of California Press.

Witkin, H. A. 1967. A Cognitive-Style Approach to Cross-Cultural Research. *International Journal of Psychology* 2: 233–250.

Witkin, H. A. and J. W. Berry 1975. Psychological Differentiation in Cross-Cultural Perspective. *Journal of Cross-Cultural Psychology* 6: 4–87.

Witkin, H. A. and D. R. Goodenough 1981. *Cognitive Styles, Essence and Origins: Field Dependence and Field Independence.* (Psychological issues 51). New York: International Universities Press.

Yuzhakov, A. A. and A. D. Mukhachev 2001. *Etnicheskoe olenevodstvo Zapadnoi Sibiri. Nenetski tip. [Ethnic reindeer herding of Western Siberia. Nenets type].* Novosibirsk: Nauchno-issledovatel'skii institut selskogo khoiastva krainego severa.

Zherebtsov, L. N. 1982. *Istoriko-kul'turnye vzaimootnosheniia komi s sosednimi narodami.* Moscow: Nauka.

Zhuravski, A. V. 1909. K samoedskomu voprosu [On the problem of Samojeds]. *Izvestiia Obshestva Izucheniia Russkogo Severa* 12: 10–28.

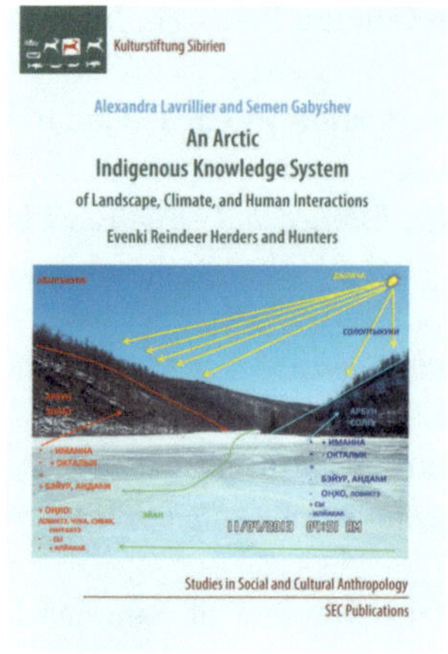

Alexandra Lavrillier and Semen Gabyshev

An Arctic Indigenous Knowledge System
of Landscape, Climate, and Human Interactions:
Evenki Reindeer Herders and Hunters

2017, Fürstenberg/Havel: Kulturstiftung Sibirien
ca. 470 pp., ca. 300 color photos, 15,5 x 22 cm
ISBN: 978-3-942883-31-3
Euro 68; paperback

Studies in Social and Cultural Anthropology
https://dh-north.org/publikationen/an-arctic-indigenous-knowledge-system-of-landscape-climate-and-human-interactions-evenki-reindeer-herders-and-hunters/en

---

Stephan Dudeck

Der Tag des Rentierzüchters:
Repräsentation indigener Lebensstile
zwischen Taigawohnplatz und Erdölstadt in Westsibirien

2013, Fürstenberg/Havel: Kulturstiftung Sibirien
351 pp., 20 color photos, 15,5 x 22 cm
ISBN: 978-3-942883-17-7
Euro 34, paperback

Studies in Social and Cultural Anthropology
https://dh-north.org/publikationen/der-tag-des-rentierzuechters/en

---

Tatiana Safonova and István Sántha

Evenki microcosm:
Visual analysis of hunter-gatherers' lifestyles
in eastern Siberia

2019, Fürstenberg/Havel: Kulturstiftung Sibirien
176 pp., 19 x 27 cm, 43 color pages
Euro 48; paperback
ISBN: 978-3-942883-36-8

Studies in Social and Cultural Anthropology
https://dh-north.org/publikationen/evenki-microcosm-visual-analysis-of-hunter-gatherers-lifestyles-in-eastern-siberia/en

---

Erich Kasten (Hg.)

Mensch und Natur in Sibirien:
Umweltwissen und nachhaltige Naturbeziehungen
in Zeiten des Klimawandels

2021, Fürstenberg/Havel: Kulturstiftung Sibirien
ca. 300 pp., 15,5 x 22 cm, ca. 80 color pages
Euro 38; paperback
ISBN: 978-3-942883-74-0

Exhibitions and Symposia
https://dh-north.org/publikationen/mensch-und-natur-in-sibirien/en

---

Erich Kasten (Hg.)

Schamanen Sibiriens
und ihr Vermächtnis

2021, Fürstenberg/Havel: Kulturstiftung Sibirien
284 pp., 15,5 x 22 cm, 47 color pages
Euro 38; paperback
ISBN: 978-3-942883-71-9

Exhibitions and Symposia
https://dh-north.org/publikationen/schamanen-sibiriens-und-ihr-vermaechtnis/en

---

Erich Kasten and Raisa Avak (eds.)

Even Tales, Songs and Worldviews
Bystrinski district
(Even / Russian / English)

2014, Fürstenberg/Havel: Kulturstiftung Sibirien
200 pp., 15,5 x 22 cm
Euro 18; paperback
ISBN: 978-3-942883-20-7

Languages and Cultures of the Russian Far East
https://dh-north.org/publikationen/even-tales-songs-and-worldviews-bystrinski-district/en

---